D0744339

Salmonopolis

Researched by
DUNCAN STACEY

Written by
SUSAN STACEY

Salmonopolis:

The Steveston Story

HARBOUR
PUBLISHING

Published by:
Harbour Publishing Co. Ltd.
P.O. Box 219
Madeira Park, B.C. V0N 2H0

Cover painting: John Horton
Cover design: Roger Handling, Glassford Design
Page design and composition: Lionel Trudel, Aspect Design

Photographs are from BCARS (BC Archives & Records Service), BCP (BC Packers), BSHS (Britannia Shipyards Heritage Society), CFC (Canadian Fishing Company), CRA (City of Richmond Archives), CVA (City of Vancouver Archives), DS (Duncan Stacey collection), GGCS (Gulf of Georgia Cannery Society), HS (Harold Steves collection), NAC (National Archives of Canada), SHS (Steveston Historical Society), UBC-JC (University of BC Library, Special Collections, Japanese Canadian collection), UFAWU (United Fishermen & Allied Workers' Union), VMM (Vancouver Maritime Museum), VPL (Vancouver Public Library).

Printed and bound in Canada

Canadian Cataloguing in Publication Data

Stacey, Duncan.
Salmonopolis

Includes bibliographical references and index.
ISBN 1-55017-110-0

1. Steveston (B.C.)--History. 2. Fisheries--British Columbia--Steveston--History. I. Stacey, Susan Lecompte, 1949- II. Title.
FC3849.S73S72 1994 971.1'33 C94-910753-0
F1089.5.S73S72 1994

Title page
Front dock of Imperial Cannery (BCP)

Opposite
Salmon gillnetters tied up at Steveston (James Crookall, CVA)

Salmonopolis:
The Steveston Story

CONTENTS

Everyone who survives a history class comes away with someone's interpretation of the past, those facts selected from humanity's vast storehouse of experience to define who and what we are as a community. Our textbooks reveal the conventional wisdom about winners and losers, good guys and bad guys, good laws and bad laws. We can look into the alternative versions offered by our libraries, bookstores and educational television stations, and if we pursue the investigation of the past, even as amateur sleuths, we soon discover that we live in an age where many groups are defining themselves by telling their stories and creating their histories. It seems that whoever has a voice has a story, and when a few voices speak together we have a history.

Steveston, a small farming and fishing community at the mouth of the Fraser River within the city of Richmond, has many voices bidding us to examine its past. For years it had a small and stable permanent population which hosted a very large and tumultuous transient population during the fishing season. It was, for a time, home to a vast diversity of humanity. Steveston has dealt with the challenges of pioneer experience, boomtown gusto and turbulence, multiculturalism and racism, rapidly changing technology, and encroachment by a larger population. From these experiences has emerged a hardy, vigorous community with a very strong sense of a heritage worth preserving.

In this book some of these voices speak again to tell us about the lives they led and the community they built. We make frequent use of quotations from written and aural records. People speak in the language and context of their times, which may require forbearance by modern sensibilities. Some memories, especially those generated in non-English speaking communities, found no verbal expression but were recorded in photographic images. The first chapter will give an overview of Steveston's evolution from an open, largely unpopulated bog to a modern fishing and tourism centre. Succeeding chapters will describe this history more vividly.

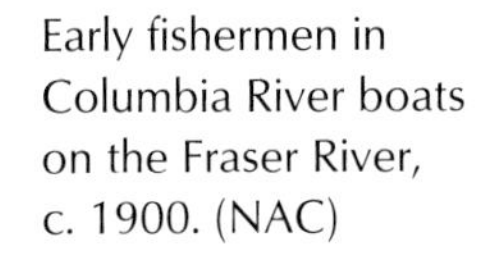

Early fishermen in Columbia River boats on the Fraser River, c. 1900. (NAC)

We are indebted to the people who found time and energy to record their memories and impressions of Steveston. This includes members of the first European families to settle on the southwest corner of Lulu Island, such as Ida Steves, Thomas Kidd and Edwin DeBeck. The book also draws on the writings of those who were just passing through, such as the journalists and missionaries who reported their Steveston experiences. We are also grateful to the people who collect and care for these treasures, including the staffs of the Richmond Archives, the Steveston Historical Society, the Gulf of Georgia Cannery Society, the Britannia Heritage Shipyard Society, the Special Collections Division of the University of British Columbia Library, the British Columbia Archives and Records Service, The City of Vancouver Archives, the Vancouver Maritime Museum, the United Church British Columbia Conference Archives, the Anglican Archives of the Diocese of New Westminster, the Salvation Army Archives, the Jewish Archives of Vancouver, the Canadian Fishing Company, B.C. Packers, the United Fishermen and Allied Workers Union, Canadian Heritage–Parks Canada–Gulf of Georgia Cannery National Historic Site, the Nikkei Internment Memorial Centre (New Denver, B.C.) and the National Archives of Canada. Marilyn Clayton, Peter Robertson, Mary Gazetas, George Brandak, Janette MacDougall, Bob Stewart, Doreen Stephens and Sheila J. Grand Scrutton were particularly helpful. Greta Cheverton spent countless hours recording the memories of Ida Steves, and David Jelliffe correlated the summaries of over one hundred interviews in the *Oral History of Richmond*, a project of the Richmond Nature Park Committee. Recognition is also due to those who have already examined these records and published their own interpretations, in particular Leslie Ross, T. Ellis Ladner, Thomas Kidd, the Richmond Nature Park Society (11851 Westminster Highway, Richmond, BC) and the individuals involved in the Richmond '79 Centennial Society Historical Sub-Committee. And finally, thanks is due to those who assisted in the creation of this effort to interpret Steveston's past: Harold Steves, Cathy Steves, Eleanor Harper, Ralph Turner, Tad Hayashi, and Jack and Doreen Deagle were especially helpful.

Cannery workers (called "cannery girls" at the time), 1940s. (CVA)

Introduction

"Take a Steveston car and ride with it until it has carried you to the cannery village by the Fraser. Your ride will take you from nearly all that is familiar, and in the end you will find yourself where the speech of five tongues meets you with a pleasant shock. The explosive gutturals of totem-faced Indians, the harsh words of wandering ex-soldiers of Sikh regiments or sailors from the navy of Nippon, and, running through the babel like song of a violin, the thin cries of the Chinese, remind you that here is the land where story books are written. Even the English has the oddness of vernacular bred in lone places far from where centralization keeps it pure.

"Here are the dream skies of romance, the shaded marionettes of many tales. Here, in bas-relief, is the essence of those hard-boiled lines we used to read in a public school geography anent the Fraser River and its salmon. In this village of unpainted buildings are the canneries and the people who work in them. Houses and canneries might be so many shacks and barns, for there is nothing to distinguish them. If you stand outside on the prairie-like land of Lulu Island and look at Steveston there is little to attract you. You will see no trace of the myriads of things that make Steveston one of the strange pictures painted by nameless waifs and tucked away in odd corners of the world."
(Garnett Weston, *British Columbia Magazine*, 1911)

The BC Electric Railway tram, bound for Steveston from Vancouver, c. 1908. (VPL 2179)

Thousands of years ago, a small island took shape in the delta region of the Fraser River, its southwest corner a point of exit into the Gulf of Georgia (officially named Strait of Georgia) and the vast Pacific Ocean. Like so many other islands of silt it gained and lost ground constantly as the river ebbed and flowed with the tides and annually sent great floods of melting ice sweeping down from the interior of the continent. There were no huge stands of forest, no plants or animals unique to this grassy bog fringed by a few deciduous trees.

The aboriginal people of the delta region – the Coast Salish – may have had permanent settlements on the island; they certainly had propriety claim to parts of it during the summer fishing and berrying seasons. All the complexities of this Native culture were as foreign as the geography to Europeans caught up in the exhilaration of the Age of Exploration. The exploits of Sir Francis Drake, Juan de Fuca and Vitus Bering fired the eagerness of merchants and their homelands to seek riches on the Pacific Coast. None of the reports of these voyages indicate that they located the mouth of the Fraser, but parts of de Fuca's account have been taken as proof that he found the Strait of Georgia.

Captain George Vancouver. (VMM)

In the 1770s Spain dispatched the explorers Juan Perez and Bodega y Quadra to the north Pacific Coast to assert its interest and to search for the coveted passage connecting Hudson Bay to the Pacific. They were followed closely by James Cook, on his third voyage of exploration to the Pacific for Britain. None of this trio was successful in finding the desired trade route, but their adventures fanned the fires of mercantile interest to white-hot heat. New waves of exploration followed in the 1780s. Most of these efforts skirted the Straits of Juan de Fuca and Georgia, thus missing the Fraser River but finding plenty of other landmarks on which to bestow the names of the explorers.

By the 1790s the rivalry for the North Pacific had narrowed to two contenders, Spain and England. Spain built a military post at Nootka Sound on the west coast of Vancouver Island, to protect its claims from foreign intrusions, and from this base sent out expeditions looking for the strait described by Juan de Fuca. On one of these voyages, in the summer of 1791, the Spanish evidently discovered the "sweet water" of the Fraser, though they seem to have missed the south arm which flows past Steveston.

The next summer the Spaniards Dionisio Galiano and Cayetano Valdez and the Englishman George Vancouver continued charting the coast in search of the river to Hudson Bay, occasionally meeting and comparing notes. Both expeditions charted the Gulf of Georgia, but by-passed the south arm of the Fraser whose shoals seemed unnavigable. George Vancouver did leave the first European name on the area near Steveston, however. He purchased some sturgeon from Natives at Point Grey and named the western edge of Steveston's island Sturgeon Banks. It was over thirty years before another European ship visited the islands in the Fraser's delta.

Britain eventually gained control of the north Pacific Coast, but as the Spanish withdrew their claims, Americans began to show an interest. Britain decided to establish a right to the territory and its fur trade by building a chain of fortifications along the coast.

While exploration by sea failed to locate the elusive waterway between the Pacific and Hudson Bay, other explorers were closing in by land. In 1793, Alexander Mackenzie reached the Pacific by way of an overland route. In 1808 Simon Fraser descended the river that bears his name, but the islands at the mouth remained uninvestigated because the interior stretches of the river were of no navigable use; there was no good reason to build a fortification at the mouth of a river that could not carry anyone beyond its first hundred miles.

By the mid-1820s, the strategy had changed. The Hudson's Bay Company (HBC), the North West Company and American fur traders had all been busy exploring the land between the Rockies and the Pacific, from Fort St. John and Fort Nelson in the north to the mouth of the Columbia in the south. In 1821 the HBC and the North West

Company merged, concentrating and consolidating their fur-trading activities in the northwest. The HBC governor, George Simpson, wanted to use the Fraser to link the Pacific and the interior, and to challenge the maritime activities of the Americans.

Acting on Simpson's instructions, in July 1827, Chief Trader James McMillan set out from Whidby Island on the sailing vessel *Cadboro* for the mouth of the Fraser to find a safe site for an HBC fort. On reaching the river, the men spent nine days at the entrance of the south arm trying to find a channel through the sandheads, during which time the north point of this arm was named Point Garry (now Garry Point) in honour of Nicholas Garry, Deputy Governor of the HBC. Eventually choosing what is now Fort Langley for the outpost, the traders named two of the upstream islands after clerks on the *Cadboro* (François Annance and George Barnston). The two large islands at the mouth remained nameless.

A map of the main arm of the Fraser River and the area's canneries, c. 1900. (Lionel Trudel, from a NAC map)

The establishment of Fort Langley may not have fulfilled all of Simpson's aims, but it did increase activity at the mouth of the Fraser. Fort Langley prospered as a fur-trading post and a marketing centre for the agricultural and fishery products of the region. In 1846 the land north of the 49th parallel, plus Vancouver Island, was granted to the British in the Oregon Boundary Treaty. By 1849 the HBC had moved the centre of its operations from Fort Vancouver on the Columbia River to Fort Victoria at the southern end of Vancouver Island. The company planned to make greater use of the Fraser as its communications link to the interior forts. Only boats engaged in company trade were allowed to use the river. That same year, however, gold was discovered in California and a great many British settlers hastened south, attracted not only by the gold fields but by the price of land.

Seeing the need to revise its settlement scheme, the HBC persuaded the British government to send the Corps of Royal Engineers to survey the new colony of British Columbia. Under the command of Colonel Richard Moody, the engineers arrived in New Westminster, the newly-established capital of British Columbia, and produced their first maps in 1859, including the island on which Steveston is situated. Covered mainly by grasses and deciduous trees, the island was desirable for agricultural settlement because it had no forests to be cleared and levelled. Thomas Kidd, an Irishman who had sought adventure in New Zealand before immigrating to British Columbia and becoming one of the island's first settlers, described his new home as follows:

Thomas Kidd, Richmond pioneer and historian. (CRA)

> On these Islands, before they were disturbed by the white man, a considerable growth of timber along or near the water courses existed. A crabapple growth along nearly all the gulf side of both islands was an outstanding feature, with a spruce tree here and there to make its outline, at a distance, among which was one known as the Point Gerry [sic] Tree and appreciated by mariners entering the Fraser River, but which became a victim to the remorseless work of the Fraser many years ago. Near the north end of this row on Sea Island a clump of spruce ended this margin of growth along the gulf side of this island.
>
> The south side of Lulu Island...was bare of timber and then about a quarter of a mile from the river was a mixed growth of spruce, cedar, hemlock, alder, some yew and on the outside cottonwood, crabapple and elderberry...With the exception of the timber growths mentioned and the peat bogs which cover nearly one third of the island, the rest of Lulu Island was covered with grass of different kinds and hardeck, with here and there some willow scattered in small patches. There were also some patches of reeds and cat-tail flags in places not so well drained as others...

> Along the water courses, where the timber grew, especially where the crabapple and willow bushes stood close to the edge, wild roses grew in great profusion and to a great height, garlanding the bushes and festooning the trees, whose beauty in June was indescribable....
>
> On the large islands there were patches of good grass, blue joint and red top, which the early settlers found very useful for hay, which could be cut and cured, after the freshets in the river were over, in late July and August. These grasses furnished food for cattle all the year round. Such was the condition of these islands before the white men began their work of reclamation.
>
> But long before that, they were for many centuries good hunting ground for the Indians as they were an ideal home for such fur bearing animals as the beaver, musk rat and mink....
>
> On these islands, deer were plentiful and some bear, the first breeding thereon, but the latter swimming across thereto in the early summer for blueberries and other fruit, generally returning to the mainland for the winter. (Thomas Kidd, *A History of Richmond Municipality*, 1927)

The sternwheeler *SS Beaver* on the Fraser River, 1890s. (VMM)

Soon after it was surveyed, the first property on Islands No. 1 and No. 2 was purchased and settled. Near the area of what is now Steveston, a Mr. Wilson of Victoria bought land in 1860 or 1861, prompting further interest in surveying the land along the south arm of the river. Wilson dyked the land and constructed some buildings, but had a difficult time finding tenants for it.

By this time British Columbia's gold rush had started. To encourage settlement of farmers on the rich delta lands to provide food for the mining districts in the interior, Governor Douglas pushed through the Pre-emption Act of 1860. This allowed farmers the right to 160 acres of unsurveyed land, provided they occupied and improved it immediately and paid the government a maximum of ten shillings an acre when it was surveyed. In order to get men and supplies to the gold fields, river transport was expanded as well. In addition to the HBC vessels *Beaver* and *Otter*, steamboats from the US began operating between Victoria and Yale, the head of navigation on the Fraser. Although these vessels passed by Island No. 1, they could not stop because there were no landing facilities.

Both New Westminster and Victoria grew rapidly, developing amenities such as the theatre and social club built by the Royal Engineers in New Westminster. One of the most popular entertainers to grace that stage was an actress from San Francisco named Lulu Sweet. When she moved on to a performance in Victoria, Colonel Moody accompanied her on the voyage and was engaged in conversation with her as their vessel passed the large island at the mouth of the south arm. When Lulu Sweet asked what it was called, Moody gallantly bestowed her name upon it, and in 1863 the former No. 1 Island appeared as Lulu Island on the charts.

During the next two decades the island gradually filled with farming families, many originally from Ireland, Scotland and England. Many had arrived in BC via Australia and the United States. Some had tried their hand in the gold-mining districts of California or the Cariboo, or in road building or other support services for the mines. Because the boggy, low-lying interior of the island was often flooded, farms, surveys, dykes and other improvements started along the perimeter. As a result, roadbuilding and settlement were delayed for many years. Transportation was most easily accomplished by water or along the dykes. Most homes were built facing the river with private wharves for their boats.

Until transportation networks evolved, the island developed several small communities rather than a single cohesive one. These included Eburne, Bridgeport, the South Arm, East Richmond and Steveston. In 1879 settlers on Lulu and Sea islands (Island No. 2) petitioned the province to become a municipality and, in November of that year, the islands became the Corporation of Richmond, later renamed The Corporation of the Township of Richmond. The Township was empowered to undertake public works such as roads, dykes and bridges, but in its early days, when cash was scarce, these transport links were primitive and Richmond remained a collection of small settlements. Even as the municipality grew, Steveston felt itself a distinct and unique element in the civic scheme of things. This separateness was supported by a ward system of municipal government, in which each community within the municipality had a representative on council to look after its interests, a system that continued until after World War Two.

Manoah Steves headed a family which moved from New Brunswick to Ontario and then to Baltimore. Steves headed west in the fall of 1877. After careful investigation he settled in the southwestern corner of Lulu Island, bringing his family out in May 1878. They dyked the land and started a dairy farm. His son, William Herbert Steves, bought the land that became the Steveston townsite, west of No. 1 Road and south of what was then known as No. 9 Road and is now Steveston Highway. Some of the names given to roads in the townsite, such as Moncton and Chatham, reflect the family's New Brunswick and Ontario connections. Other families established farms in the vicinity and agricultural enterprises were soon underway. It was not Steveston's fate, however, to become a small farming community similar to so many others in the province. Instead, the life cycles of the settlers of Steveston were deeply influenced by the life cycle of *Oncorhynchus nerka*, the fabled sockeye salmon.

Fort Langley had a fish salting operation that proved so successful, several private salteries were established in New Westminster. Some salteries experimented with salmon canning technology, and a new and extremely prosperous industry was born. As competition increased, canners moved downriver to be closer to the fish as they entered the Fraser. Steveston was as far downriver as the canners could go and many large canneries were soon constructed, attracting large numbers of labourers during the summer processing season. Accommodations were built for the workers and merchants arrived to supply the growing needs of the fishing and farming communities.

Initially, Natives supplied the bulk of the canneries' labour force. The men fished while the women knitted and repaired nets and worked in the canneries. The Natives came from various villages along the coast, following a seasonal work pattern that brought them to the canneries in the early summer and then to the hop fields of the Fraser Valley and Washington state in early fall. Chinese men, many of whom had been labourers on the railway, were also employed as cannery workers, but almost never as fishermen. They were hired by the canners through the services of a Chinese contractor who negotiated wages, living arrangements and working arrangements, and acted as intermediary and translator between employer and employee. Living in bunkhouses near the canneries during the summer, they worked the rest of the year in mining or logging camps or on the railway.

Entertainer Lulu Sweet, for whom Lulu Island (the former Island No. 1) was named. (CRA)

Japanese fishermen, encountering poor working opportunities in their native land, heard descriptions of the salmon fishing on the Fraser and decided to try their luck. At first they were migrant workers, arriving in Steveston in springtime and returning to Japan in the fall if they could afford to do so. But the prosperous fishery led many to become permanent residents of Canada. At first they worked only as fishermen; later they also became boatbuilders. The Japanese lived in their own bunkhouses, with a house-boss acting as liaison with the canners for a percentage of each resident's earnings. When the fishermen began to relocate their wives and families to Steveston, married housing was built along the boardwalks. Japanese women found work in the canneries.

Europeans were involved in the fishery as cannery owners, office workers, tradesmen, overseers, fishermen, cannery workers and boat builders. They came largely from areas with fishing traditions such as the British Isles, Scandinavia, Italy, Greece and Yugoslavia. Fishermen from the US also migrated to the Fraser River in the early years before laws were passed limiting licences to British subjects. These workers also moved to other seasonal occupations after the fishing. A member of the Steves family, a child in the first years of the twentieth century, remembers:

> **This town varied according to the seasons. An awful lot of people came in the summer for fishing, and all these canneries took quite a crew. By 1906 there were as many as 10,000 people in Steveston. You'd go down there on a Saturday night when the fishing was closed – they had eight-foot board sidewalks through the town and I guarantee you'd likely have to walk down the road there was so many people standing on the walks you couldn't get through 'em, all different nationalities. And quite a few Indians, too. They'd come down from their reservations from way up north. See, they'd work in the canneries. They employed a lot of people in those days because they didn't have that modern machinery and they had to cut the heads off and take the innards out and everything, all by hand. (Harold Steves, Sr., in *Steveston Recollected*)**

The canners were independent businessmen with experience in numerous enterprises. Generally white and middle class, they saw canning as a relatively sound investment, with a seemingly inexhaustible and reliable resource in its early years and a market among the working class of England. They arranged financing through commission houses in Victoria, imported technology from the canning industries on the Sacramento and Columbia rivers, and formed an association to promote their own interests. Various labour-saving technologies were adopted or invented over the years until, by 1912, the canning line was virtually completely automated. A number of canning companies consolidated into larger and more powerful entities.

The Steveston cannery docks were a harbour for sailing vessels from all over the world. Hotels lined the boardwalks, and by 1891 Steveston had a nine-hundred-seat opera house, along with dance halls, drinking and gambling establishments, social clubs, ice cream parlours and other recreational outlets. There was brief excitement

Sailing ships loading canned salmon from Steveston canneries, 1898. (VPL 1856)

at the discovery of gas in Steveston. High-tech experts arrived from Texas before the prospects were found to be commercially unfeasible. Farmers and European-born residents of Steveston found their commercial and social centre along Moncton Street. The Chinese community gathered at the foot of No. 1 Road, and as the Japanese became more prosperous, they too started shops and services.

The local newspaper boasted that Steveston, a.k.a. Salmonopolis, would become the centre of the lower mainland area. It didn't. A series of disastrous fires and poor salmon runs in the early twentieth century saw the end of the dream. Canneries began to close and sailing ships found other harbours. By World War One, Steveston had passed its glory days. Still, it was in no danger of becoming a ghost town. Some canneries remained, the hotels stayed open, and businesses, schools and churches grew. The Wild West atmosphere gave way to a more family-oriented community. An interurban train service connected Steveston with Vancouver. There were plans to make the harbour a major West Coast port facility.

World War One took away much of Steveston's manpower, but most of the soldiers returned and life went on. Farming and fishing continued to provide livelihoods and the community continued to develop. A ferry dock was built at the end of 7th Avenue and a ferry service to Vancouver Island began. Rum runners using Steveston's harbour facilities during American prohibition supplied lots of excitement. Roads and bridges were built and upgraded to improve communication with the rest of the lower mainland. Sports and social activities provided entertainment, and when the Depression of the 1930s swept the West Coast, Steveston's residents were not as desperate as those in other communities. Money was scarce but food was available and public works helped with other expenses.

Like so many other Canadians, Stevestonites were more concerned with the economy than with politics and were caught by surprise with the arrival of World War Two. The community set up a civil defence fund and organized a full-scale war effort. Unlike the rest of Canada, Steveston was hard hit when Japan bombed Pearl Harbor and most of the community's residents were suddenly under suspicion as potential saboteurs. In April 1942, the federal government, with the support of the Richmond municipal council, removed all Japanese persons from the coast, loading them onto trains and sending them to the BC interior and to southern Alberta. This grim turn of events, inflicted on Steveston by the fears of those outside the community, probably changed the town's history more than anything that had happened before.

After the war, and the gradual return of some Japanese residents, Steveston and the rest of Richmond experienced a period of great growth, still sustained by farming and fishing, but augmented by the baby boom and the recognition that Richmond was a favourable place to raise a family – property was both cheap and close to Vancouver. Agricultural land disappeared as subdivisions

Tashme, a Japanese Canadian internment camp near Hope, BC, 1940s. (UBC-JC VIII-86)

developed, and the old ward system was abandoned. Councillors now represented the municipality at large. A town planning bylaw was enacted to oversee development. With the new residents came new services, including sewers, roads, bridges, the Deas Island Tunnel, expanded medical/dental facilities, shops, theatres, social services and recreational facilities. The airport developed, new churches were built, and there was an increase in community groups and service organizations. The latter included the Steveston Community Society, a vehicle for Steveston residents to get involved in civic affairs and area park development, the building of the Martial Arts Centre, and the annual Salmon Festival.

Steveston's unique heritage has created a community that remembers and celebrates the varied elements of its history. Those traditions feed a strong community that has managed to retain its identity and history in the face of many overwhelming forces.

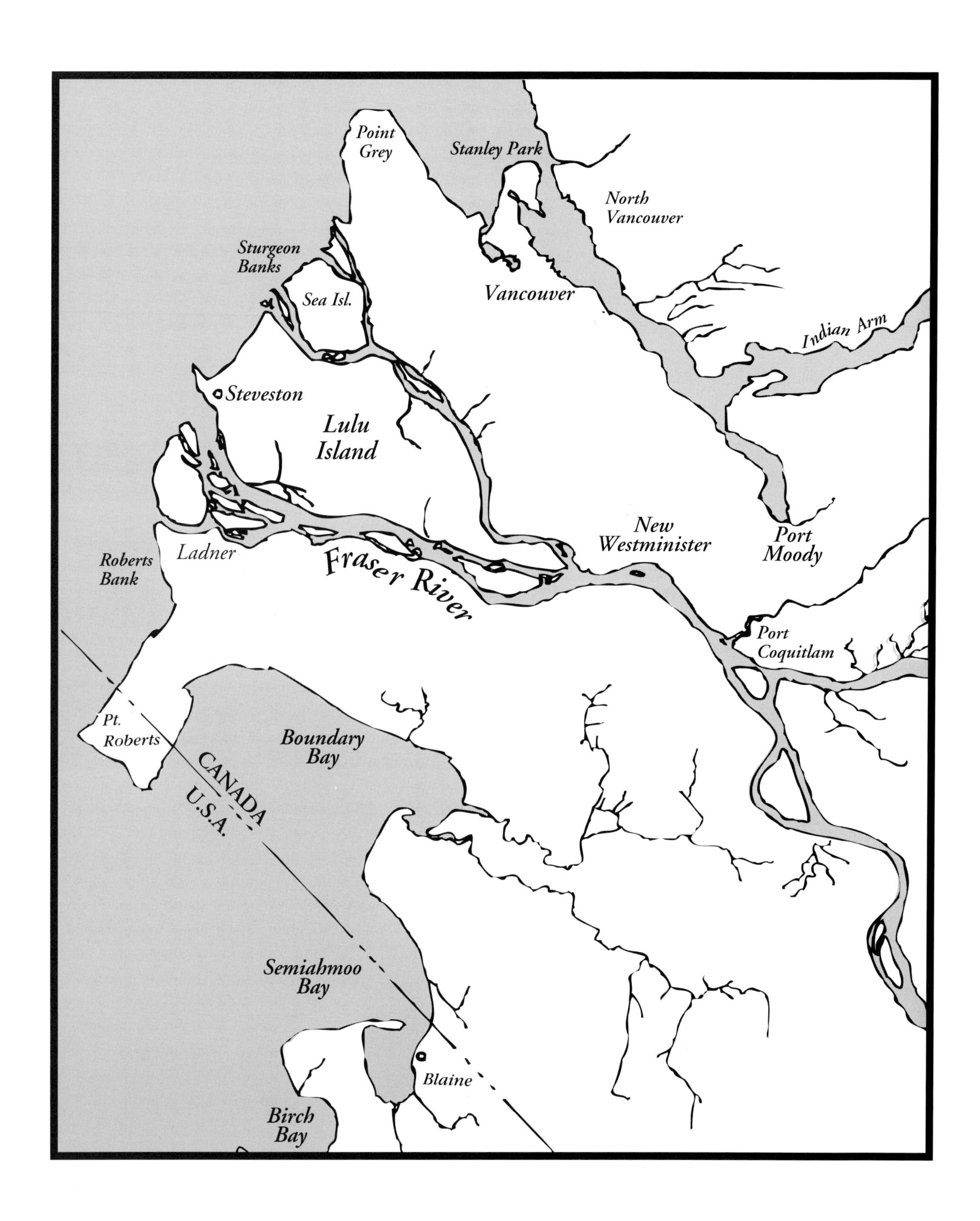
Point Grey
Stanley Park
North Vancouver
Sturgeon Banks
Sea Isl.
Vancouver
Indian Arm
Steveston
Lulu Island
New Westminister
Port Moody
Roberts Bank
Ladner
Fraser River
Port Coquitlam
Pt. Roberts
Boundary Bay
CANADA
U.S.A.
Semiahmoo Bay
Blaine
Birch Bay

Chapter One:
The Farming Community: Early Years

We are going to invite your attention for a moment to…a field with verdure clad, and plains of emerald, whose paths drop fatness; where the cow looks for the milk-maid, and the lambs troop together for play…the large extent of Island prairie in this district is most valuable, not only as grass and stock farms, but for all green crops, fruits and even cereals…we shall shortly have a large and thriving agricultural settlement between [New Westminster] and the Gulf of Georgia, and all these beautiful and fertile islands will, ere long, be covered with waving corn and blooming orchards. (*British Columbian*, Summer 1862)

Hard at work in the fields of Steveston, c. 1900. (HS)

When Manoah Steves arrived in British Columbia in 1877, he found the area around Lulu Island similar to that of his native New Brunswick, but more promising for his young family. He decided to settle in the southwest corner and bought some lots from A. E. Sharpe. His wife Martha and their six children, ranging in age from two to twenty years, arrived the following May, the first large family in the area. They took advantage of the transcontinental railroad in the United States to make their way to their new home. One of his daughters, Ida, later recalled the journey.

Original home of William Herbert Steves and family, c. 1890. (HS)

There were two ways to travel from Baltimore west: express train or immigrant train. The immigrant train was slow and the people took their own bedding with them. There was one special car at that time for the Americans, and all good quality people travelled in that car. They stopped a day or two in Baltimore, then left Cambridge on the 6th of May by B. and O. and reached Lulu Island on the 24th of May. When the conductor called "Harrisburg, Pittsburg, and the West" we had to change to a slow train. They changed trains at Chicago and had to take a bus to get to the other station. This was not long after the big fire, and they saw the black buildings that hadn't burned, just were smoke blackened. The Central Pacific Railway took them from Chicago to Frisco. There were many cars of immigrants from Central Europe. The women all wore kerchiefs over their heads. They didn't mingle with the others because they couldn't speak English. The Steves had a nice carload travelling together, singsongs in the evenings, and a nice social time together. They had to get off the train just past Sacramento. The river was in flood. Ripe cherries were on the trees, and the water was clear up to the branches. They went to Frisco in small boats. It was three days from Frisco to Victoria by steamer. The family landed in Victoria at 6 in the morning. They had to hurry around and get their baggage checked and put on the Enterprise which left at 8 A.M. The Enterprise was a side wheeler which ran to Westminster. (Ida Steeves, *City of Richmond Archives,* Biography Files)

The sidewheeler *Enterprise*, which brought the Steves family on the last leg of their journey to their new home. The photo was taken shortly after the *Enterprise* was rammed by the *R.P. Rithet* near Ten Mile Point, Victoria, in 1885. It sank in Cadboro Bay. (VMM 7790)

When Mrs. Steves asked to disembark at the southwest shore of Lulu Island, near the wild crabapple trees and the lone fir tree, the captain of the *Enterprise* protested there wasn't even a wharf. Mrs. Steves insisted, and on reaching their destination the family was lifted into a row boat by sailors and set down on a huge log. "The reeds covering the flats were over six feet high and although the father…had walked out on the log to meet them he couldn't be seen until the row boat came along the side of the log." (*Richmond Review*, June 4, 1958)

Steves's property, which included a small shack built on stilts, ran along the western shore almost to Garry Point and was partly submerged at high tide. There was some question as to the exact boundaries of his land; he later bought more property from owners in Victoria, increasing his holdings to 400 acres. Once reunited, the family settled into the cabin and began discovering the peculiarities of their new home.

> Mr. Sharpe had a lot of pigs that ran out on the flats. There was some kind of grass that had a very nourishing root, something like "chuffus" grown in the gardens in the east. It was all gone within three or four years. The pigs rooted up every bit of grass. The pigs at South Arm fattened on skunk cabbage roots. They sold these to the Chinamen. No one else would eat the pork.
>
> Sharpe had a barn built, and a two room cabin and some other buildings. It was the only house here at the time. When Sharpe went to Vancouver he told Steves he could live in his house, so we camped there until the house was built.
>
> Wild roses grew up over the tops of crabapple trees, too high to reach. Also there were gooseberries as big as cherries along the beach. When we went to pick the gooseberries, some Indian women at Garry Point told us to go home because these were the Indians' 'Olillie'.
>
> Sharpe's cow barn and his house were near 6th Avenue. He had a dyke when the Steves came here first. Drainage was very poor. It had to drain from the Gulf. Ditches would fill in with grass during the winter.
>
> There were a lot of wild cattle on the prairies here, when the Steves family arrived. You could go out on the prairies and see a band of wild cattle [descendants of livestock lost or abandoned by farmers on other parts of the island] out there amongst the hardhack and grass. They would all stand in a bunch as soon as they saw a person, and then if you made a step towards them they would turn and run as fast as they could go. When you walked down to the river in summer evenings, some of the wild cattle would be out on the beaches come down to drink, maybe four or five big bulls, out there bellowing and scrapping. (Ida Steeves)

Manoah's eldest son, William Herbert Steves, bought land north of what is now Chatham Street in 1880 and then bought waterfront property south of his first purchase around 1887.

> When Herbert found out that most of our homestead was tidal flats, and the lines were all wrong, he went to Victoria and he bought the Steveston townsite quarter. At that time the ownership from No. 2 Road down was Londons', Gerrards', Englishes', the townsite belonging to Herbert Steves, Bell Irving. When Blairs came in they bought above Londons'.
>
> Manoah bought the piece just east of him from Charlie Hughes (between Steveston Highway and Williams up to No. 1 Road). Chinamen who had come up from San Francisco cleared part of this land for the use of it. Carvall had a piece along the waterfront north of Manoah's…Mother had property at the corner, and another piece along No. 9

Scotch Pond and dyke at the west end of Steveston, 1950s. The Steves residence is in the upper right corner. (CFC)

The Steves family home, built in 1890 at the corner of No. 1 Road and Chatham Street. (CRA)

Road. Aunt Alice owned clear down to the barn. . .In the East the wife always owned one third of the property. (Ida Steeves)

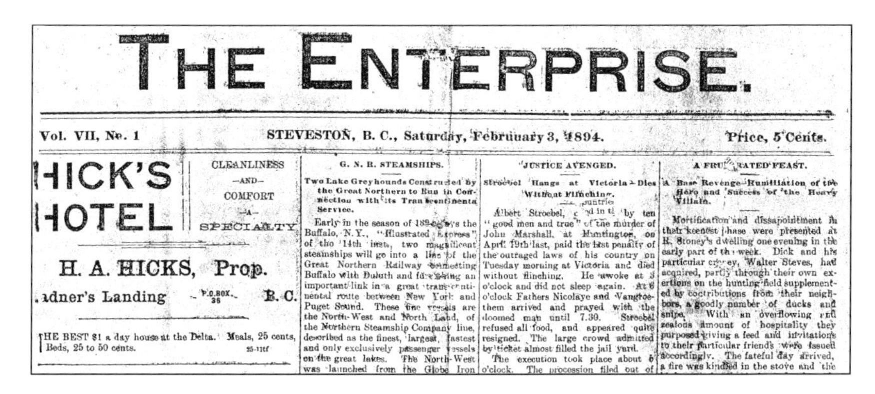

THE ENTERPRISE.

Vol. VII, No. 1 | STEVESTON, B. C., Saturday, February 3, 1894. | Price, 5 Cents.

HICK'S HOTEL

CLEANLINESS —AND— COMFORT —A— SPECIALTY.

H. A. HICKS, Prop.

Ladner's Landing - P.O.BOX. 36 - B. C.

THE BEST $1 a day house at the Delta. Meals, 25 cents, Beds, 25 to 50 cents. 25-11tf

G. N. R. STEAMSHIPS.

Two Lake Greyhounds Constructed by the Great Northern to Run in Connection with its Transcontinental Service.

Early in the season of 189[illegible] says the Buffalo, N.Y., "Illustrated Express" of the 14th inst., two magnificent steamships will go into a line of the Great Northern Railway connecting Buffalo with Duluth and forming an important link in a great transcontinental route between New York and Puget Sound. These fine vessels are the North-West and North Land, of the Northern Steamship Company line, described as the finest, largest, fastest and only exclusively passenger vessels on the great lakes. The North-West was launched from the Globe Iron

JUSTICE AVENGED.

Stroebel Hangs at Victoria—Dies Without Flinching.

Albert Stroebel, c[illegible] by ten "good men and true" of the murder of John Marshall, at Huntington, on April 19th last, paid the last penalty of the outraged laws of his country on Tuesday morning at Victoria and died without flinching. He awoke at 3 o'clock and did not sleep again. At 6 o'clock Fathers Nicolaye and Vangroe-them arrived and prayed with the doomed man until 7.30. Stroebel refused all food, and appeared quite resigned. The large crowd admitted by ticket almost filled the jail yard.

The execution took place about 8 o'clock. The procession filed out of

A FRUSTRATED FEAST.

A Base Revenge—Humiliation of the Hero and Success of the Heavy Villain.

Mortification and dissapointment in their keenest phase were presented at B. Stoney's dwelling one evening in the early part of the week. Dick and his particular crony, Walter Steves, had acquired, partly through their own exertions on the hunting field supplemented by cotributions from their neighbors, a goodly number of ducks and snipe. With an overflowing and zealous amount of hospitality they purposed giving a feed and invitations to their particular friends were issued accordingly. The fateful day arrived, a fire was kindled in the stove and the

Herbert Steves laid out a portion of his land in town lots, and started a newspaper, *The Steveston Enterprise*, as a vehicle for promoting investment in the new community. According to Thomas Kidd, this publication was "so absorbed in the business of booming the place of its birth it scarcely recognized the existence of any other place or paper. When it did refer to Vancouver it always called it 'Steveston's outer port,' in a jocular way of course." He also built an "opera house," a large building used for public gatherings of many kinds. Thus began the community first known as the Town of Steves, becoming Steveston in the summer of 1889.

How this place got started was, W.H. Steves, along with Manoah Steves, set out to make a town. They figured they could even compete with Vancouver, they started ahead of Vancouver really, and they figured they were going to make a big city here. But they sunk too much money into it and they lost most of it. The Sun Life Company had the mortgages, they were the ones that finally sold the land. About 1891 they had a brochure out. You could get a big lot for $150 or a smaller lot for $100. They held a big auction sale and hired boats to bring people from New Westminster, Victoria, Nanaimo, and I guess from somewhere out in Moodyville. Had a big gala day, barbecued an ox and so forth, and sold the lots. (Harold Steves, Sr.)

The efforts of Wm. Herbert Steves to build a town at the mouth of the Fraser River to rival the young city of Vancouver were not as successful as he had hoped for. That Vancouver had the start of about five years, with a transcontinental railway terminating there, and one of the best harbours in the world, and many other advantages to support it, did not seem to lessen his hope and faith in his undertaking, for he pursued his purpose with great activity and tenacity, called to his

The Steveston Opera House, on the southwest corner of Second Avenue and Chatham Street, served as church, meeting place, community hall, performance hall, dance hall, etc. Built in 1890 by Herbert Steves, it was eventually torn down and the lumber was recycled to build a house. (CVA P665)

aid all the auxiliaries within his power of persuasion, and even started a newspaper called the "Steveston Enterprise" to advertise and extend his propaganda. That it was to become the centre of the salmon industry of the Fraser, was apparent, but this was not sufficient to inspire a general belief that it would become any more than a large village. The progress that it did make before he died was not enough to recoup him for the great outlay which he had made and some think that his financial loss and disappointment helped to shorten his life. (Thomas Kidd)

Even though he wasn't the first owner of land at Steveston, Manoah Steves was its first true European settler and as such had to build better dykes to keep his farmland from flooding.

At that time the homesteads were all close to the waterfronts and no public works of any kind existed. The dykes that had been built to fence the water out were all undertakings of the individuals owning the land on the water courses and without any joint action other than that every owner was glad to have a neighbour come in and join dykes with him.

In the earliest efforts some dyked all around a small area, while others built a front dyke and ran smaller wing dykes to prevent ordinary tides from overflowing. Indeed very few at that time were sufficiently dyked to prevent some flooding in the winter time; nor was a little flooding in the winter considered a very serious matter. (Thomas Kidd)

The attractiveness of a low-lying island with few trees to clear was offset by its vulnerability to flooding. Because the interior of the island was so boggy, most of the early farms were built on its perimeter, and were often in danger in flood season. Although early settlers built their own dykes to protect themselves, some found the effort and expense excessive and looked for less exposed land. The municipal government, which in its first few years could not finance dyking projects, started road construction across the island to enable settlement farther from the river. These early roads were covered with gravel. Settlers owning land adjoining them were allowed to cut ditches and use the dirt to grade the roads, with the council paying ten cents for every cubic yard of earth graded.

During this year, petitions for the construction of road ditches to be dug poured into the council, all of which could not be granted but the councils of those early days were making the limited revenue go as far as possible in this work. Many of the petitioners were not so very much concerned about getting roads as they were to get outlets for the drainage of their land, which incidentally came to them by the construction of these road ditches, and the non-resident owners were amongst the most eager for this work to be carried on, while the resident owners, who were, almost without exception, located on the watercourses and had established drainage for themselves to a large extent, found the additional drainage from the road ditches very helpful and led to all being eager to have such work done. (Thomas Kidd)

Ditch water was controlled by floodboxes that prevented saltwater from entering the ditches at high tide. The facts of life in early Steveston were that dykes could fail (especially if muskrats had colonized them), that floodboxes weren't entirely reliable, and that ditches caved in when one could least afford the time to rebuild. Farmers devoted as much of their time and resources as they could to dyke maintenance and put a ten cent bounty on muskrats. But after two major floods, in 1894 (a deluge that changed the course of the river) and 1905, landowners decided to establish a local dyking commission. This body had the authority to request tools, men and materials to build or repair dykes, as well as the power to assess owners or occupiers of land for any expenses incurred in construction or repair work. Steveston came under the authority of the West Lulu Island Dyking District Commission, which brought in a floating dredge to dig up sand and gravel to build up the dyke.

The early farmers put up hand dykes. They hired Chinese labour to dig them up with spades. Sometimes the water came right over. The highest tide when the dyke was broken never came right into the house, just up to the door. Sometimes we had to use a rowboat to fetch firewood. Father invited some people from Steveston to come to our place for Christmas dinner. They couldn't get there. The tide came in all over Steveston.

About 1908 they had a big steam shovel on a scow, self propelled. That's when they dug the big ditches and made the dykes bigger. (Ida Steeves)

Even with the best drainage efforts, however, early life in Steveston required a sturdy pair of gumboots, and it was customary to carry a pair of slippers to change into when visiting. The ditches also provided entertainment and early play areas for children, providing swimming opportunities, habitat for muskrats and other creatures, and nature studies of various other kinds.

There was a swimming hole in the ditch near the Steves house where the boys used to swim naked. One day Winnie and Jessie Steves walked up towards this swimming hole and saw a lot of naked bodies jumping off the dyke into the ditch. They turned around to go back, giggling. Allan, Jessie's brother said later, "You should have kept on coming. We were all under anyway by then." (Ida Steeves)

Waterskiing on planks in a flooded field. (BSHS)

While the residents of Steveston waged a seemingly unending battle to keep excessive amounts of river water out of their farmlands, they were also plagued with the problem of finding a reliable source of uncontaminated drinking water. Earliest residents carried river water in whatever containers were available to their homes, where it was boiled and filtered through charcoal. Rain barrels were another means of collection. Sometimes river water was allowed to sit in barrels to let the suspended matter settle out. The water supply might also be reduced in winter. When ice was still frozen at the headwaters, the tides from the Gulf of Georgia flowed farther up river, making the water saltier. Ditch water was too alkaline for drinking and not very desirable anyway, since many outhouses were built over ditches. Whatever the local source, it was universally agreed to be foul-tasting stuff. When the Marpole Bridge was opened, Stevestonites could get water from the North Arm artesian well at 25 cents per bucket. When the railway came to Steveston, the cans that transported milk from Richmond to Vancouver returned full of water. Cannery boats stopping in at New Westminster brought water on their return trip. The municipal council let several contracts for artesian drilling in various spots on Lulu Island, but the quest for a local fresh water supply was futile. In 1909 Richmond made an agreement with New Westminster to have water from the New Westminster reservoir supplied to Richmond's boundary. The next year Richmond began on installing a state-of-the-art water main system. The system was plagued with cracks, leaks and other problems until the municipality decided, in 1930, to join the Greater Vancouver Water Board and take advantage of the pooled resources of several communities.

The earliest Steveston settlers were almost exclusively men; if they had families they tended to stay in New Westminster or Victoria until a cabin had been built. Then they loaded their household goods on some form of river transport, landed at the closest wharf—perhaps the one at Deas Cannery on the south side of the South Arm—and completed the journey to their new home by scow.

....the Kidd and Lee cabin [was] the centre point of meeting in those early bachelor days of the settlement. [Lee's] knowledge of and facility in cooking, added to a natural sociability of character, made that cabin a place for Sunday and holiday gatherings, but there

were no spongers among those early settlers, so that partnership lost nothing by the rude hospitality given there as long as it was confined to those settlers only...Mrs. John Green in those early days of the settlement lived in Victoria most of the time, which put her husband in the bachelor class during her absence and made him eligible for those gatherings.

There were no dress suits on these occasions and indeed any effort towards "city style," was almost resented. Gum boots and getting about in skiffs and canoes were not conducive to giving expression to high ideals in that direction. Even the necessary trips to the Royal City [New Westminster], where they were called mud-flatters, did not incite them much to special dressing, for a 12 or 13 mile pull in a row boat was rather damaging to white collars, especially to paper ones which were much in fashion in those days. (Thomas Kidd)

A resident of Lulu Island says that fully 50% of the whole population of the Island is composed of marriageable bachelors. This is an opportunity, for enterprising damsels, to secure a good husband and a good home. The settlement is thriving well and there is lots of room for more settlers. The land is excellent and capable of producing enormous crops and all that is lacking now is more desirable settlers. (*The Vancouver News*, March 8, 1887)

Railway construction made emigration from eastern Canada easier, but the island continued to attract a largely British population and those accustomed to lowland farming.

About this time the construction of the C.P.R. was a growing certainty and settlers began to come into B.C., of which these islands got some, but as a general rule the Ontario immigrants did not like the idea of having a dyke to depend on to keep them from being flooded...Indeed, one that could or would not handle a spade, an axe and an oar in those days was in the wrong place on these islands. (Kidd)

As time and money permitted cabins were replaced by houses, but farm life continued to provide some peculiar tribulations.

By fall they had the hay all raked and baled and were going to use the money from the sale of the hay to build a house. A wind smashed the barn, so they had to stay in the old house and build a barn instead.

Two men wait for the BC Electric Railway tram. (CRA)

The original house was built in 1878. They cut away a few crabapple trees, and Herbert Kirkland did most of the building. The first year we were pretty hard up for money. We sowed a small patch of oats and it didn't come up till fall. We used it all for feed. We got some trees from Ernie Hutchison...we visited Hector McDonald "up the slough" and got sweet apple, half a dozen at first, all grafted on to the crabapple...One of the trees was a golden russet, planted seventy-five years ago in 1863, and still alive and bearing today.

The first house was quite small, with a large living area downstairs and a sleeping area upstairs. A lean-to kitchen was added on the south end. On the east side of the house a lean-to bedroom about twenty feet long and eight feet wide was added. One night after Martha Steves had gone to bed, Alice and Ida heard a crackling sound, and found the shingles around the stovepipe had caught fire. The children were asleep upstairs and Mother in the northwest end of the house. This had happened before likely, but this was the worst. Alice stood at the bottom of a ladder they kept there up to the roof, and handed buckets of water from the rain barrel up to Ida who threw the water onto the roof and put the fire out. Afterwards Mother wanted to know what they were doing out there. This kitchen was added on sometime after the house was built...The living room window was on the west, facing the dyke. One time there was such a storm, spray covered the window, with a coating of salt so thick they couldn't see through it.

There was a space between the bedroom lean-to and the kitchen lean-to. Ida, then about thirteen or fourteen, and her younger brother Walter decided to fill this in for a pantry. They were not too successful because of the different angles of lean-tos. Rats could easily get in. Walter laid shingles and Ida nailed them. One time when Walter was home and sitting in the living room, Ida reached up on a pantry shelf and grabbed a rat, a full grown rat, by the neck. She held onto it and called Walter to come and kill the rat, and Walter did. The rats used to gnaw holes and get into the milk house. Ida had a stack of heavy window glass she had brought up from the store. The rats would hide behind this, and Ida would squash them against the wall and kill them...This home was just to the east of the present Steves home on the same site. It was finally torn down in the early 1950s. Many of the original orchard trees remain. (Ida Steeves)

Nearly all the farms soon had private wharves for transporting their families and goods and eventually for transporting their produce to market.

Captain Middleton had a crew of three and a Chinese cook. He owned his own schooner and would lease it to the government. The men rowed in with the tide to Steves for fresh vegetables and fruits, and went out with the tide, during the summer. Father planted potatoes on March 1st. These were sold to Onderdonk. Sternwheel boats loaded vegetables and took them as far as Yale, but they only ran in the summertime.

Two Steves boys were going out to the Enterprise to take some cabbages up to the New Westminster market and Sharpe [also known as "Big" Sharpe, reputed to be at least six feet, six inches tall] was going to New Westminster at the same time. Herbert and Sharpe were manoeuvring to get the boat close to the Enterprise right out in the channel, to unload the little boat, which was piled high with cabbages, and they hit the steamer when they moved around, and the little boat tipped right over completely bottom up. The two boys grabbed the boat and got up on the bottom of the boat, and sat there, but Mr. Sharpe didn't know how to swim, and he just grabbed hold of the side of the boat and Herbert got hold of him to hold him up. Then the Enterprise put out her small boat, and gathered up the boys and Mr. Sharpe and then the cabbage which were floating around. Herbert had no trouble finding some other clothes to put on, but they hunted all around all the men on the boat for something big enough for Sharpe to wear. They hadn't planned on giants going up there. When the next New Westminster paper came out, it told about how Sharpe walked out in the river to stop the Enterprise with sacks of cabbages. (Ida Steeves)

After dykes, homes and wharves were built, the first settlers turned to the difficult task of transforming their properties into farms. This was a slow process, since the newly drained land took a long time to become solid enough to support the weight of horses and wagons. The earliest settlers used teams of two or three yoke of oxen and a twenty- or twenty-four-inch breaking plough to start reclaiming their land. When the first crops were produced they were hauled to storage areas on sleighs. Thomas Kidd, referring to himself as "the writer,"

does not remember when he saw the first angle or earth worms on these lands, but he knows it was some years before they appeared in the land after it was dyked to do the great work for the improvement of the soil which, Charles Darwin pointed out, is the result of their borings. And of course, moles did not appear until there was some appearance of safety that they would not be drowned out.

Owners of oxen customarily transported their animals from New Westminster, plowed south Richmond fields, then walked them around the perimeter of the island to a scow that took them back across the North Arm to the mainland.

Hay could be grown on these lands, and "it need hardly be said that both primitive tools and methods were used in this work, scythes and hayforks. And in stacking, the haycocks were carried on two poles, and when too far away from the stack were drawn by a pole and a rope which in this case Green's pony was found useful" (Thomas Kidd). As the road ditches drained more land, larger areas were put under cultivation.

After dyking, draining, ploughing and cultivating his land, the elder Steves started a dairy operation, first with local stock and then with imported purebred Holstein cattle, which became a prize-winning herd.

The first cows were rented. We had them before we bought any. We rented them on shares and kept the new stock. Rosie's mother was a red cow, just common stock. Cherry was a calf Herbert bought from Harry Sexsmith and raised. Ida got two pigs, and raised them and sold one for $25, and she bought Cherry from Herbert. Later she traded her offspring for purebred Holsteins.

Mother had three little pigs, Baby, Spot and Sandy who came around to the door all the time. She got the pigs from Sharpe and they liked to be scratched with a stick. They would come in the house if we let them.

They sent along a bull from Victoria, and the bull accidentally got killed. It was a shorthorn, and we weren't anxious to have that bull over here anyway.

They got the first cow from Sam Brighouse. Mother wanted a nice quiet cow that a woman could milk, but father got this real wild thing. She got out among the wild cattle and ran with them. In the end we made beef out of her, sold her to some butcher or something. She had a calf which got out among the wild cattle the same summer. A little steer calf got into our garden, a lot smaller than the one that got out. Sweet had a lot of wild cattle then. Father told him he could have the wild calf, because it had the same wild nature as its mother, and Father got the little calf that had come through the fence. We named it "Pinkie" and had no other cows for quite a long while.

Father went back to Chatham a couple of times to dispose of the property there. He bought three horses and some cattle. He wanted to get purebred stock, so he went down to Washington and bought two heifers and one bull. Black Princess was the name of one of the heifers. He went back to Ontario to buy a carload of cattle, and came back with two carloads. There was no place prepared for them. They were fine cattle, a lot of them imported from Holland...The man told him he could sell bulls up the Fraser Valley and set his own price on them. But farmers had not been educated to appreciate purebred stock. (Ida Steeves)

Steves later extended his interest in pedigreed stock to Suffolk Punch horses and Berkshire pigs.

One of the first bylaws to apply to Steveston prohibited hogs from running at large. Other livestock control measures soon followed. By 1889 there was a municipal pound for wandering swine, and a keeper in charge of collecting and caring for them until their fines were paid or they were destroyed or sold at auction.

Herbert Steves also prospered as a farmer and seed merchant. In addition to the dairy operation, the Steves land and surrounding farms produced hay, mangles, fruit, vegetable crops and grapes. The hired help needed to supplement the family efforts was provided by local Native and Chinese workers, and by newly arrived Europeans earning money to buy their own property.

A lot of Indians were on the threshing gang. The threshing machine was brought from Ladner on a scow, and it always came to Steves place first. The Indians worked on the straw stack. The gang was here only a week the first time, but they had some later grain, and it rained and rained and the crew stayed nearly two weeks then. There was no blower on the machine and men had to pitch the straw away. They hired teams for hauling, bought meat for the meals, using from twenty to thirty pounds of meat a day. (Ida Steeves)

For a while the Steves had an employee named John Scott.

Herbert or someone, had tied the boat down by the Indian ranch after coming from New Westminster or Ladner. Father said there was a wind coming up and we shouldn't leave the boat there. He told John Scott to go down in the evening and take the boat and put in the little slough [by the Imperial Cannery]. John was very willing to do anything at all so he went down to the Indian ranch and found the boat and took two ropes off the stake and rowed up the slough. He thought it was rather heavy, and when he went to untie the ropes he found two sturgeon on one rope. The Indians

A picnic on No. 5 Road, possibly part of a day of blueberry picking. Left to right: Jean Steves, Leleah Wescott, Jessie Steves. (HS)

had tied their sturgeon to the same stake as the boat. John Scott took both ropes with him. When he came back father told him that it would not do to take the Indians sturgeon. He had always treated Indians fairly. He said, "Instead of milking cows in the morning, go down to the Indians and tell them about the sturgeon. Go to Charlie's house, go to the door and tell them what happened." Mother told him, "Do not try to talk any Chinook." Emily, Charlie's wife, was a half breed and could speak almost better English than John Scott.

John went down, but he didn't follow instructions. He stood out on the walk and called out, "I want to klatawa with Emily." He said this over and over again many times. "I want to klatawa with Charlie's wife." Nobody opened the door or made a sound. Everything was very quiet. Finally someone poked his head out and asked what he wanted, so he told them about the sturgeon. The Indian said, "Why didn't you tell us that first time." Emily told the Steves about it afterwards. The Indians didn't know what to make of this strange man coming there and wanting to run away with Charlie's wife. (Ida Steeves)

In addition to providing cash income, farm produce fed the family as much as possible. In the early years buying anything from stores was expensive and required a long trip into New Westminster.

The first two or three years we used old whiskey bottles for preserving. The first fall we came we didn't have any fruit. We went out and picked a lot of wild crabapple, cooked them up and strained them, and put them in whiskey bottles. We made jam because sugar was too scarce. It was sixteen to eighteen cents a pound. We also made preserves out of wild gooseberries. Father got in five barrels of flour at a time. They never used anything smaller than a ten cent piece until the small nickel came into use. If the price was fifty-three they paid fifty-five, if it was fifty-two they paid fifty. (Ida Steeves)

Some of the local produce was sold to other Steveston residents.

The Steves farm sold butter to cannery cookhouses...Alice churned the butter about three times a week, making about 100 pounds a week. Miss Murray in the lighthouse cookhouse got butter from Mrs. Steves. Once she got butter from Mrs. Savage on Westham Island, but said her butter did not keep as well as Mrs. Steves' butter. Mr. Monroe, of the Phoenix Cannery, was very disappointed when he was dropped from the list, when Mrs. Steves could no longer supply him. He bought butter from Mrs. Smith on River Road on the North Arm. Mrs. Smith also peddled her butter from a rig in Vancouver. Mrs. Trites who ran the Commercial Hotel bought butter from Steves. She liked lots of salt in it. Around 1908 they stopped making and selling butter and sold milk instead. (Ida Steeves)

Most of the produce and dairy products from the Steveston area, however, was taken to market in New Westminster, where it was sometimes trans-shipped to Victoria or Nanaimo. It was no small task to do this, involving "the need of skiffs and expenditure of muscle, to get to New Westminster, and, indeed, until the roads were got into good condition, most of the farmers on the south side of Lulu Island went to New Westminster to do their selling and buying." (Thomas Kidd)

Transport for the first settlers was most easily accomplished by boat. The first homes were built facing the river and had private docks. The dykes could be used for foot traffic, but not for horses or wagons. The first municipal council meetings were held at farmers' homes and had to be scheduled for moonlit nights so that council members, who couldn't spare any daylight hours for meetings, could see to navigate their boats. Settlers travelled by rowboat and canoe, but the favoured mode of family transport was the flat-bottomed scow. This was such a staple of the early transport system that the bridges over the sloughs were designed with consideration of the scow's dimensions.

> **[First concert, around 1880:] They went in the boat, went out on the tide across the Gulf to the North Arm of the river...They came to a building, probably a church on the right hand as they walked up from the river...When they started for home after the concert they had to go around to the main channel because the tide was out. They walked the logs in, went around to Garry Point and tied up the boat. John Scott and Herbert rowed the boat. Ida, Alice and Josephine had benches like it was in a church. (Ida Steeves)**

There was a great deal of small boat travel to New Westminster, the major commercial centre in the early decades, but until the first cannery was built in Steveston in the early 1880s, there wasn't a wharf facility large enough for a steamboat to land. Although English and Company, owners of the cannery, were generous with the use of their dock, there were no roads leading to the cannery, which limited its usefulness to farmers. When a large public dock was built at London's Landing at the south end of No. 2 Road in the mid-1880s, residents of Steveston finally had a place for the larger steamboats on the Victoria–New Westminster run to land with freight and mail.

> **The Enterprise went to New Westminster twice a week, went up one day and came back the next. Irvine was the first company that owned the Princess Louise. It was also called the Yosemite and the Olympia Teaser. The Otter was a sternwheeler. Sidewheelers did not go up the river to Hope...People from Ladner dressed up to go to New Westminster. People from Steveston didn't dress up for the trip. Steveston people were plain country people. Somehow or other the Ladner people looked as if they had just come from the hairdresser's, and their clothes were just so...Sailing ships took six months from the Old Country around the horn to Steveston and New Westminster. (Ida Steeves)**

This was a significant project, linking Steveston directly with Victoria and New Westminster and providing a local mail depot, another service residents had formerly received from New Westminster, and then from Ladner.

> **We used to get our mail at Ladner. Ladner people came by boatloads to visit in the**

Salmon gillnet fleet at Garry Point, Steveston, c. 1900. (DS)

summer, and anyone who came brought the mail with them. Sometimes we went across in the boat to fetch it. (Ida Steeves)

The dock also enabled a daily steamboat run aboard the *Telephone* to carry people, freight and mail between Steveston and New Westminster by way of Westham Island and Ladner. The demand for river transport was so heavy that another, larger boat, the *Edgar*, was soon built. One farming family also constructed a dock for shipment of grain, hay and machinery to Vancouver, Victoria and Nanaimo. By the turn of the century the farmers of Steveston were well connected to their markets by river transport. They welcomed the advent of the Canadian Pacific Railway (CPR) as a means of increasing the province's population and thereby their markets, and were encouraged to improve their holdings and equipment to meet the projected needs.

In the very early days of European settlement, even before they could land at Steveston, the river boats did provide one service to the Steveston farmers.

The whistle of the boat running between Victoria and New Westminster, which was twice a week in the summer and once a week in the winter, was the only aid the settlers had for the greater part of the year to escape this, and when this was missing, as it was in the winters when the ice in the river prevented the boat from getting up to Westminster, they were as badly off as Robinson Crusoe on his island in respect to a record of time. (Thomas Kidd)

Within Steveston the earliest mode of transport was by horse. Ida Steeves remembered that "Everyone rode horseback before they started the canneries." It was not always a trouble-free way to get around.

Herbert bought horses from up country. Joe had to break them in. Going to church one Sunday [at No. 2 Road and the dyke] he hitched one of these wild horses, the best of the lot, up with a quiet gentle mare, just a pony. They were driving along Number 9 Road [Steveston Highway] just about where Railway crosses it...A Chinaman was walking along carrying a bundle of some sort of red blankets on a stick across his shoulder, or maybe two bundles, one at each end of the stick. The horse made a jump. The little horse fell down. The wagon didn't turn over, but the wild horse ran into the barbed wire fence...Joe took one of the seats out of the wagon and threw it over to stand on because it was very muddy and wet and he had his Sunday shoes on. He quieted the horses down as well as he could, got them turned around and straight up on the road again, fixed the seat, and climbed back in...Another time the same team of horses was pulling a sleigh. Snow was deep that year, and Thompson was preaching. The Smiths' sleigh was right ahead of them. The whole Smith family was in the sleigh with a lantern on the back of it. This same wild horse took fright at the lantern, and jumped off the side of the road into deep snow, and broke loose from the harness and left the little mare there, and ran away. Walter and Joe went sprawling headlong into the snow over the front of the sleigh. (Ida Steeves)

Roads were another transport link desired by Steveston's early residents, though roadbuilding and maintenance were expensive and some feared the additional land taxes they would incur. Nevertheless, nine roads were gazetted for Lulu Island in 1881. The first one built was No. 2 Road in 1883, followed by No. 1 Road and No. 3 Road, all north–south routes. Later construction produced No. 4 Road, also north–south, No. 9 Road (later Steveston Highway) running east–west, No. 6 Road, No. 5 Road, No. 8 Road (all north–south), No. 19 Road (later Westminster Highway) and the River Road, as well as several roads in Steveston townsite. Roadbeds were large stones and rocks. Residents earned 40 cents per hour hauling this material from the quarries. Ocean-

Jessie Thompson (standing) with Blossom, a horse, and Violet Thompson (second from left on horse) and Mary Thompson (right), 1914. (CRA)

Second Avenue, looking north from near the dyke, 1898. The Steveston Methodist Church, built in 1894, can be seen in the distance. The Richmond Hotel is at extreme left. (CVA)

going ships bound for Steveston sometimes carried roadmaking rock for ballast. Smaller stones topped several layers of rock, all of which was steam rollered into the ground. Some roads were planked initially because that was a faster construction technique. The wharf area in Steveston, Moncton Street, 2nd Avenue and 4th Avenue in Steveston, and Steveston Highway from 4th Avenue to No. 3 Road, and a portion of No. 3 Road and No. 5 Road all were built this way.

No. 2 Road was the first road built right across the island. No. 1 Road was built long from Steveston for half a mile or so, and No. 3 Road was mostly corduroy [around 1890s]. In Steveston the plank roads were eight feet wide. Indian women sat on the edge of the sidewalk and hung their feet over the edge. No. 1 Road was gravelled at that time. Sailing ships coming to pick up the salmon carried ballast of rocks. These rocks were dumped on the road. No. 9 Road was like soup, deep mud, soft after the fall rains. There were eventually planks on the south side on No. 9 Road. The north side just had the prairie, not even a ditch. Only one ditch was made and the ground thrown up to make the road and the planks put on top. No. 1 Road was dug by hand with ditches on each side. When the dredge came through later they filled in the ditch on the west side of No. 1 Road. (Ida Steeves)

Because most employers paid Chinese labourers lower wages than their European counterparts, the Steveston municipal government—like many in the province—became caught up in the fear that the poorly paid Chinese worker would undermine the local economy. In 1885 the council moved that only white labourers could be employed on public works projects except where they would not work for less than 25 percent more than the lowest Chinese tender. Further, a contractor could not sublet his contract to Chinese crews.

From an early date municipal leaders were campaigning for bridges across the North Arm of the river to connect the island with the wagon road (now Granville Street) into Vancouver. After years of negotiations with provincial, federal and Vancouver city authorities, and much financial hand-wringing, bridges connecting Lulu Island with the road into Vancouver were opened in 1890. Residents of Steveston could take their produce to the growing Vancouver market and do their shopping there more easily. There was, however, the added financial burden of maintaining these structures. To prevent excessive wear and tear, speed limits were imposed—

Principal retail shops along Second Avenue, 1908. (CVA Out P676, N283)

allowing a horse to trot instead of walk across a bridge earned its owner a fine.

By this time there were some thriving dairy operations in Steveston and the bridges enabled dairy farmers to transport unpasteurized milk to Vancouver by wagon. It was a very long trip, however, and if there were delays or the weather was too warm, the milk spoiled, to the dismay of Vancouver buyers and health authorities. Dairy producers were equally unhappy with the Vancouver merchants who watered down their milk or added preservatives. This discontent led to the formation of a milk commission in Vancouver and eventually to the Fraser Valley Milk Producers' Association in 1913. The milk commission inspected dairies and graded their milk. Other, less perishable farm produce also made its way to market over the wagon road to Vancouver until the arrival of the railway in 1902.

With roads developing, there was a demand for stage lines and livery stables. Steveston had several, including a stage line run by Humphrey Trites from the southeast corner of Chatham and 3rd Avenue, A.H. Wescott's "Sockeye Stables" at the northwest corner of the same intersection, and "Billy" Steeves's "Palace Livery Stage and Sales Stables," which made a daily Steveston–Vancouver run. The trip cost fifty cents. Walter "Billy" Steeves, a distant cousin of Manoah Steves's family, spelled his name with an extra *e*. He married Manoah's daughter Ida, whose reminiscences, told in the third person, are a major part of this book.

Passengers took the stage where the Steveston Hotel now stands, or were picked up on the way. It was a two hour trip to Vancouver, so round trip was all day. Woodward's had a place in the centre of the floor where the ladies could dry off wet skirts on a rainy day. One day in 1890 Aunt Alice and Ida and Joe Steves drove over to Vancouver. It had rained all night before and it was very muddy on Hastings Street. The stage left early in the morning, 7:00 or 7:30. Men had to get out and walk up the hill on Granville. The horses could not haul the full stage up the hill.

Once we went on a picnic. Walter Steeves lent his stage for the transportation. Alice and Ida sat on the back seat. Now on the trip to Vancouver this back seat was always reserved for

Chinese and Indians. Ida's husband was horrified that his wife and sister-in-law had ridden there, after he had lent them the stage. (Ida Steeves)

In December 1895, during the run home from Vancouver, a tree, its roots loosened by gravel-hauling operations in South Vancouver, was uprooted by the wind and fell between the horses and the stage. Steeves was killed instantly, but the Chinese passengers on the rear seat and the horses were unharmed.

Ida married Walter in 1891. He ran a stage line to Vancouver. He would go round in the evening in Steveston and up the dyke to see if people wanted anything from Vancouver. He left Steveston early in the morning, would have lunch in Vancouver and be home for supper.

He was killed on December 23, 1895 by a tree falling across the stage. That day the minister, Mr. Green, had made arrangements to go on the stage, and just as he came to get on, someone else came along with his horse and buggy and invited the minister to ride home with him. Otherwise he would probably have been killed too. The horses were not hurt at all. The Chinese on the back seat were not hurt. Only the driver was killed. The South Vancouver workmen had taken so much gravel out from under the tree that the heavy wind blew the tree down. Ida got a handsome settlement from them. (Ida Steeves)

After 1912 many new roads were built. The number of automobiles and other motorized vehicles grew rapidly. BC motorists followed the British practice of driving on the left-hand side of the road until January 1, 1922, when the change to right-hand drive was made.

Once roads and bridges were in place, attention turned to building an interurban railway. The first proposals were presented in 1893 and centred on transporting canned salmon from Steveston to Vancouver. The scheme didn't get past the planning stage because of the seasonal and unpredictable nature of the canning trade and because other interurban projects were not prospering at that time. Nine years later, however, in 1902, the CPR built a light rail line to Richmond and started twice-daily runs of the Vancouver and Lulu Island Railway. In 1905 the British Columbia Electric Railway Company took over operation of what was now known as the "Sockeye Limited." Canners did not immediately use the rail line. They shipped their product by boat because it was more direct, less expensive and required less handling from cannery wharf to ship's hold. Transport by train meant getting the cases to the one rail stop in Steveston, then transferring to other trains heading East. The rail line did service the canneries by transporting people to and from work, and eventually canners switched their shipping practices when rail transport proved more practical. The farmers of Steveston, however, made immediate and enthusiastic use of the line to ship their products into Vancouver, helped no doubt by the special rates the railway made available to them. As farmers prospered, the railway moved its Vancouver terminus to a location closer to the farm and dairy warehouses. In 1906 the line switched from steam to electric power. For the next fifty years, trains ran the fifteen miles between Vancouver and Steveston on a half-hour basis, a service that is ingrained in the memory of everyone along the route.

The Steveston—Vancouver Royal Mail stage, c. 1900. (CRA 1984-17-12)

Lulu Island eventually had eighteen interurban stations: Tucks, Bridgeport, Sexsmith, Cambie, Alexandra, Lansdowne, Ferndale, Garden City, Brighouse, Lulu, Riverside, Blundell, Francis, Woodwards, Cottages, Branscombe, the Y and Steveston. The line ran up Railway Avenue to Granville, east along Granville to Garden City, and across a trestle bridge into Vancouver. Later a bus service connected the Steveston station with passengers arriving from Ladner at Woodward's Landing.

Children took the train to school, and Richmond residents used it to get to and from work. It was popular on weekend evenings to ride to Vancouver theatres, and on New Year's Eve Vancouverites used the train to get to the Steveston Opera House. But although the interurban made it easier for Stevestonites to transport themselves and their goods, it decreased the commercial importance of their own town by shifting the focus for shipping and marketing to Vancouver.

In 1914 the Canadian National Railway (CNR) built a line across Lulu Island from Queensborough into Steveston, with stations at Ewan Avenue, Ewan's Landing, Woodward's Landing and Steveston (at Moncton). The line was used for two years until peat fires destroyed many of its pilings, trestles and roadbed. After numerous protests, the railway was reactivated in 1916, but poor roadbed conditions closed it down again the next year and it remained unused for fifteen years.

The Sockeye Express, an interurban tram that ran between Vancouver and Steveston until 1958. (VPL 2182)

As Steveston grew into a prosperous farming and fishing centre, plans were drawn up for a major seaport. Canners discovered they could save a lot of money by loading their cases of salmon directly onto sailing ships from their cannery wharves, rather than shipping them to agents in Victoria for distribution. For many years, Steveston's harbour was frequented by sailing ships from around the world, loading canned salmon for British and Pacific Rim markets. In 1912 Charles Pretty devised a scheme for transforming the deep, wide entrance to the Fraser into a harbour and distribution centre for Steveston's fishing operations and other commerce. Pretty's endeavour proved to be Steveston's last failing grasp at big city status. French financiers were behind the plan, but their support vanished when World War One began. Vancouver and Prince Rupert went on to become the province's major deep-sea harbours. While nothing on the scale of Vancouver's waterfront ever materialized in Steveston, it remains the nation's largest fishing port.

Communications systems connecting Steveston's residents with each other and with surrounding communities developed quickly. In the earliest days of settlement, residents relied on New Westminster and Ladner for picking up and mailing their letters while in town on other business. Mail was later delivered to the wharf at Phoenix Cannery, and a store built on London's Landing eventually became the postal depot. In 1890, a post office was established in Steveston itself. Mail initially arrived by boat, later by road, but by 1919 air-mail delivery across the Rockies was in its infancy and the low-lying terrain of Lulu Island was perfect for landing fields.

Steveston had its own early newspaper, *The Steveston Enterprise*, which kept residents informed of community activities while actively promoting investment opportunities. It ceased publication in November 1894. Stevestonites also kept abreast of things with the New Westminster weekly paper and the *Graphic*, the *Illustrated London News*, *The World* and the *Point Grey Gazette*. Magazines were sometimes exchanged at social gatherings; when the Steves family went to a concert on the North Arm of the river in 1880, Ida Steeves reported, "there were men from the logging camp on Point Grey there...The men from this camp sent over magazines."

The Richmond Hotel, built in 1890 on the west side of Second Avenue between Moncton Street and the dyke. (CVA C92-0727-135)

In 1891 the New Westminster and Burrard Inlet Telephone Company installed Steveston's first telephone in J.C. Forlong's store. When an incoming call was received, a messenger set off to find the person for whom it was intended. Phoenix Cannery hooked up to the line, and in 1898 a telephone was installed in the police office. By 1905 a municipal-wide system was planned, and in 1912 there were twelve subscribers in the Steveston area. By 1922 Steveston had one of the two exchanges in Richmond, and several dozen customers.

Electricity came to Steveston when the interurban tram line was electrified in 1905. The line supplied the canneries with lighting and some machinery operations about the same time. But many houses were lit by gas as a result of the short-lived mining boom on Lulu Island. When a water-drilling operation in Eburne brought up traces of gold, the island was staked off into mining claims and several companies were formed. Further exploration led to reports of large reserves of natural gas in 1891. Vancouver papers carried stories that a lighted match tossed into bubbling puddles caused a flash like gunpowder.

The Walker Emporium, a well-known Steveston store, c. 1900. (SHS)

Dykeside saloon in Steveston, c. 1900. (CRA 1978-37-15)

This led to hopes of coal and oil discoveries, and in 1904 the Steveston Land and Oil Company was formed. The firm bought land along Broadway, brought in engineers and machinery from Texas, and proceeded to explore eagerly. No large reserves were ever found and the engineers were plagued by a fine silt which clogged the machinery. The project was abandoned in 1906, but the small reserve that produced the inspiring bubbles yielded enough gas to light Steveston households for many years. Heat was supplied mainly by firewood collected along the river or by coal delivered from Nanaimo.

Feeling the need of an organization, Richmond farmers formed the Agricultural Society in 1890 and Steveston farmers were the proud hosts of the first Richmond Agricultural Show in 1892. Generally, however, Steveston residents preferred the bigger fairs in New Westminster and Vancouver, from which they often carried away prizes.

By the late 1880s the farmers of Steveston were reaping the rewards of their earlier hard work. Prosperity was expressed in various ways. Cabins and small houses were enlarged or replaced, and pioneer life was softened. Ida Steeves recalled that "about this time Steves bought rose bushes from rose growers in Pennsylvania back east, paid no duty. They were Dingee and Cannard, softwood cuttings less than a year old when received. One is still left, a cutting from one of the originals." The residents had several mercantile operations, centred on Moncton Street, to serve their needs. Most of the homes and farms were located within three blocks of this commercial and social centre. In 1888, J.C. Forlong bought some lots in the Steveston townsite and started the first general store. Within a few years Stevestonites could also shop for hardware and general supplies at stores owned by Mr. Rubinowitz, Mr. Branscombe, Mr. Sisson or Mr. Petersky. Eventually there was the Steveston Meat Market, fish markets, Hunt's Store (later the Walker Emporium) and a seed store owned by W.H. Steves. The Hepworth Building had a drug store, the landing at the foot of No. 2 Road had London's store and post office, and townspeople enjoyed culinary treats at the Steveston Restaurant and Maggie Quinn's ice-cream parlour. The town had machine shops, barber shops and poolrooms. Steveston residents also counted on regular visits from door-to-door salesmen offering a great variety of clothing and household items. There was a Northern Bank (later the Royal Bank) for financial services, and a theatre and opera house for pleasure.

James Forlong had the first store in Steveston, built sometime between 1885 and 1888. The Forlong and Steves Seed store was on No. 1 Road in 1888. At that time No. 1 Road was built as far as Chatham Street. Beyond that it was just as it was thrown up with the spades full of clay...Rubinowitz's had a store built right over the water beside the dyke, across from Branscombe's first store...Walker bought out Hunt's store and added to it. He wanted cash for everything...The store was...northeast corner of Moncton and 2nd. Mrs. Anno, a Japanese lady who looked after dry goods, worked there. Conway, who was very English, worked there too, then later he had his own store on the southwest corner of Moncton and No. 1 Road...London's store, run by Bill London, a general store and post office. At the back was a big dining room and big kitchen. Bill London was the postmaster until he went to California because of TB...This store was at the dyke end of No. 2 Road...Maggie Quinn had an ice cream parlour, beside the drug store on 2nd Avenue...This store later became the first Steveston school, church services were also held there...We got patterns for the latest styles from Harper's Bazaar. Mother sewed up yards of print inside covers for extra bedding. She paid about four or five cents a yard for it. The Jews had opened a store in the community and sold print...[The Steves family] brought feather beds with them, and when the pillows and beds got too old, they sold them to the Chinamen and made new ones. (Ida Steeves)

Two children at Steveston drugs, C. 1920. The notes on the photograph identify them as George and Francis, possibly George Mackey and Francis McDonald.

If not the booming metropolis of W.H. Steves's dreams, Steveston was a thriving community by the turn of the century with a reasonable optimism for the future. Canneries were springing up along its waterfront, fuelling the local economy. The workforce they attracted in the summertime challenged the town to strengthen its social structures.

Children outside the original Steveston Elementary School, 1908. (CRA 1978-1-23)

Chapter Two:

The Farming Community Matures

Unidentified man and Kanaka boy, identified in photograph notes as "Charlie," playing a guitar and banjo. (CRA 1987-104-21)

Steveston, in the Electoral District of New Westminster and municipality of Richmond, is situated on Lulu Island, near the mouth of the Fraser River. There is a daily mail service, excepting Sundays, and all boats running to New Westminster call there. Stages run twice a day to Vancouver, a distance of 12 miles. There are three salmon canneries, a first-class hotel, four general stores, seed store, wagon shop, furniture factory, butcher shop, large opera house, two livery stables and music store, besides private residences. A half-mile race track has been laid out, and the British Columbia Jockey Club has arranged to take it over, as it is considered to be the best track on the Pacific Coast. A new wharf has just been completed, having 26 feet depth of water at low water, and is large enough to accommodate any vessel. There are a number of very fine farms in the immediate vicinity....From its situation and shipping facilities Steveston will develop into a town of some importance. At present the water supply comes from the Fraser River, but the residents are negotiating with an Eastern firm to bore Artesian wells...A large number of lots have been bought there on building conditions, and even now, owing to its being the headquarters of the fishing operations, it is a very busy and thriving little place. Many of the fishermen live in house boats, which they move from place to place, but as most of the fishing is done at the mouth of the river, this is the most convenient place for them to live. In the summer they fish for salmon, and in the winter for salt water fish, finding a ready market in New Westminster and Vancouver. (*Williams British Columbia Directory*, 1891)

What a cosmopolitan town Steveston is; and yet in its peculiar features there is but one Steveston in the world. For two months in the year all nationalities gather here, so that your missionary meets with Jew, Greek and barbarian, in the literal sense of the term. The other Saturday evening, listening to the Salvation Army's "open air," there was heard the ringing of the auction bell of a sale going on a few rods away; nearby a street phonograph produced the latest music—all lost in the babble of the passing crowd, made up of fishermen in from the arduous toil of a week on Georgia's Gulf, and the conglomeration of Indian, Asiatic and European—the human life of the cannery poured forth to indulge in pleasures suited to various tastes. (Rev. William J. Stone, *Missionary Bulletin*, 1905)

By 1897 there were over a dozen canneries in Steveston. To accommodate the thousands of workers in the fishing and canning season, running from May to October, canners built rows of bunkhouses. Steveston's entrepreneurs constructed hotels and entertainment facilities for the seasonal workers and looked forward to prosperous weekends. In response to all this activity, a variety of social institutions appeared to serve the needs of the community.

The fisherman's week started at 6:00 p.m. on a Sunday night when a gunshot signalled the end of the close period.

> Sunday nights at 6 o'clock, we used to call it the 6 o'clock gun'd go off. They had somebody standing with a little rocket gun every so far on the river, you see. One down here on the point, another one over at Terra Nova. You could see the smoke in the air as the rocket went off. That was the signal to start fishing. Now in those days it was all rowboats and sailboats and you could hear those corks [cedar floats] going over the end of the boats for miles. No gas noise or anything like that, all quiet. All quiet and the gun went off and then you hear the roar of these floats going over the backs of the boats. (Harold Steves, Sr.)

A gin bottle, made about 1890. (Lionel Trudel photo)

The week ended midday Saturday (in later years at 6:00 p.m. on Friday), when the fishery was closed to allow some fish to escape upriver to spawn. Yet all was not quiet when the fishing gear was put away. The fishermen hit the streets of Steveston to spend their leisure time and considerable sums of money on alcohol, gambling and prostitutes.

> The gaming houses during the night are crowded. The hard working fishermen part easily with their gains. The large amount of earnings intoxicates the toilers with joy. This season the average gains of each was $200. I heard of one man who had lost every cent of his wages in a Chinese gaming house. I saw dollars scattered about, with the greatest freedom. People seemed to think they had so many that they might spend any number and have plenty left. Alas, that means for many a poor fellow, an empty purse and the old pant's pocket still to put it in, at the close of the season. (D. S. Moore, *Methodist Recorder,* 1899)

> But the worst of all is the dreadfully immoral condition of things about Steveston. Many nights you could not sleep on account of drunken Indians, and more degraded white men carousing around with poor, deluded women. In the day time, towards the last of the season, you could hardly walk the streets without meeting white men staggering along drunk, and using shameful language. One fellow comes along with his face all bleeding. Another poor fool, with 200 fish in his boat, selling them at 20 cents, no sooner gets them sold than he is off to one of the holes where liquor is sold, and is soon rid of the money, and in a drunken quarrel has his head split open. Two Japs, with their desire after liquor, have a quarrel with a bartender, and both get their heads cut, and the doctor is called in to sew up the wounds. (Rev. T. Crosby, *Methodist Recorder,* 1899)

> Steveston is a pretty quiet spot till Saturday night comes, when the sailor boys from ships lying here do a royal parade about 11 p.m. and rouse the neighborhood. The whisky shops are, of course, doing a great business with the Stevedore's men, who, at 35¢ per hour, can make sufficient money to have a "great time" every evening.... (John Deaville, *Methodist Recorder,* 1899)

A bottle of whiskey was a staple in most pioneer homes as a medicine and relaxant. Although there were no places one could buy it in early Steveston, stage drivers would bring it in from Vancouver and neighbours could usually be relied upon to have a stock available in case of emergency.

In the early years the temperance movement was successful in restricting the consumption of alcoholic beverages.

> Delta municipality allowed liquor, but not Richmond. There was a better element of people settled in Richmond because of that. The reeve and councillors were a better class of men. (Ida Steeves)
>
> The Good Temperance Lodge, a secret society here in the early days, used to meet in the school at the corner of No. 2 Road. They had a lodge over at Richmond, and they had started one in Steveston. Mr. and Mrs. Kidd, and the Smiths and most of the people belonged to it. They had "open meetings" in

The London Hotel, an early Steveston watering hole, c. 1908. (VPL 2184)

Interior of local drinking establishment, c. 1900. Note the spittoon on the floor.

the Opera House..."Open Lodge" took the form of entertainment or social evening. It was a secret society with a password, sentinel and guard to keep out non-members...They served something to eat, probably sandwiches and cake. We used a new shingle for a plate, and threw it away afterwards...If we used the schoolhouse on 4th Avenue, we had to bring our own lanterns or lamps. (Ida Steeves)

The Bulls were Americans, strong temperance people, especially Mrs. Bull. They wouldn't even sing "God Save the Queen" because they said the Queen was not a strict teetotaller. (Ida Steeves)

This prohibition was strongly enforced in Steveston during the fishing season. Stores advertised "Temperance Beer" but bootlegging operations abounded, and Steveston's early police records report unceasing and largely unsuccessful efforts to curtail the trade. In 1895 the municipal council granted approval for the incorporation of the Steveston Club. In July the Club opened its doors under the authority of the BC Municipal Act, as a place where its members, "whose objects of association are mutual recreation and improvement," could buy and consume liquor. Under the act the club paid an annual licence fee ($100) and its members had annual dues ($10). Spirituous or fermented liquors could be sold to members only, and the club's operations were monitored by the Municipal Licensing Commission. The Steveston Club was immensely popular and many similar organizations soon followed. It became clear that "mutual recreation and improvement" were not the prime motives for joining, and Steveston was soon notorious for its clubs. Steveston's citizens decided that rather than continue with this farce, they would have hotels licensed to serve liquor. Over half the population signed a petition to that effect in 1896. In June the licensing body considered the applications of six hotel owners, but turned them all down on a technicality. After a twelve-month waiting period, the hotel owners made sure all the paperwork was done properly and all received licences in 1897, ending prohibition in Steveston. The bars were extremely popular and profitable, and the clubs soon disappeared.

However,

> The writer would not purposely leave the impression on those who read this that before this time the people of Steveston were without opportunities to get other beverages than those concocted in the kitchen, for at this time there were several stages to and from Vancouver whose drivers were very obliging in buying and carrying parcels for their customers. The creation of clubs lessened to some extent the business of the stages, but between them supplied a demand of which they were much relieved by the granting of licenses to the hotels of Steveston...(Thomas Kidd)

> In Steveston at that time there were about 20 bars. All the men were single. Germans, Italians, Russians, all single men in temporary houses, so they all drank. The bars had small, round tables, like beer parlours today. Beer, whiskey. The beer was cheap, only 5 cents for a big glass. People drank like fish. (Asamatsu Murakami, in *Steveston Recollected*)

The bars were one of the few places men could socialize during their leisure hours.

> As men met acquaintances of former years, they fraternized. Except for rooms in the living quarters, which were often disarrayed, with unmade beds, and lit by only coal oil lamps, the barroom offered the only accommodation for socializing.
>
> There were very occasional fund-raising church concerts. These, however, radiated their own distinct atmosphere. They were quite unlike the barroom evenings, which offered the familiar singsong, with another drink reviving the gathering when the tempo slowed down.
>
> Men at the canneries were not degraded or drunkards or dissolutes, and there was no lowering of dignity or esteem by spending time in the barroom. Many of the cannery hands were highly honourable men, but all of them experienced loneliness. Without a home and without even a friend's home to visit, they turned to the saloon for their only diversion. (T. Ellis Ladner, *Above the Sandheads*)

There were also unlicensed establishments in Steveston that did a booming business selling alcohol to Natives and minors, who were both prohibited by law from possessing it.

> Steveston is a hard place. Sabbath desecration prevails, and traffic in liquor is carried on to a fearful extent. As there was no church in this place, services were conducted in the open air, or in the "Opera" (dance house), and proved very unsatisfactory. A year ago, in the fishing season, a friend from England, who takes great interest in mission work, visited Steveston, saw the people and the need of a building. This summer, just as the Indians began to settle there to be ready for the fishing, he cabled me to buy two lots for building and church, and that he would send a draft for the same...Besides Indians representing nearly every tribe in British Columbia and white men from many nations, Japanese and Chinese attended the church and listened to the Word on the streets. The Gospel seed is being scattered broadcast...Only a week ago—right here—James McRoery, a Christian white man, went from church on Sunday night to his own house on his little farm. On Monday morning he was found murdered in his bed. A drunken

Two unidentified men sitting on a sidewalk with, according to notes on the photograph, "Mackey's black dog." (CRA 1987-104-20)

Indian was arrested, and confessed the crime. He said: "I drank the whiskey; then I began to think I would go in and kill him, and I did so."...The white man who sold him the liquor gets only six months. The Indian goes to the scaffold. (Rev. A.E. Green, *The Missionary Outlook,* 1895)

Canadian prohibition, as a wartime restriction, was in effect from 1916 to 1921. While the hotels couldn't sell it, alcohol was still available from the local doctor who could issue a prescription for Jamaican rum. Patients spent $2.50 for the rum and $1 for the permit to buy it. Stevestonites also made their own beer and wine, and the local police were often occupied tracking down illegal stills.

During Prohibition in the US, rum runners made use of Fraser River harbours. Residents walking along desolate areas of the dyke might stumble upon a cache of whiskey waiting to be picked up. Dairy farmers made extra money shipping milk to Bellingham with a container of whiskey in the milk can, and fishermen, who could buy a bottle of whiskey for $5, took it aboard their vessels in hopes of meeting an American boat on the fishing grounds and selling it for $15.

In addition to the pleasures of drinking, a fisherman might indulge in a bit of gambling on the weekend. By 1905 Steveston had become so well known as a gambling centre that a council member hired a private detective to spy on the town. The detective reported there were seventeen or eighteen gambling houses wide open to the public, offering games such as fantan, poker, chuck-a-luck and blackjack, with the blackjack played so crookedly no one could possibly win. The gambling houses were patronized by Chinese, Japanese and European men, and by some women who were secretly admitted through the back doors. Only one of these places sold liquor, but it was readily available elsewhere. He also reported that prostitutes were at work in houses behind the gambling dens.

The fish were unusually late in making an appearance this year, and in consequence there were abundant opportunities for the Indians indulging in all kinds of vice. Gambling and

Hoisting a pint at a Steveston bar, c. 1899. (CVA Out P679, N286)

drinking was the order of the day when we reached the camps. Thousands of Indians were gathered at Steveston. Illicit whiskey-selling was rampant. As soon as Bro. Baker and myself landed there, one Saturday evening, we were met by several Indians, complaining bitterly of these whiskey dives. They implored us to urge the authorities to stop it, so many of the women having commenced drinking. In one den, where I endeavoured to get some of the young women out, I was grossly insulted, and this by men who called themselves white men. In an adjoining room I could hear the filthy language of others, mixed with the low laughter and chinook of some young Indian women more or less under the influence of liquor. I believe the municipality of Richmond is to blame for all the vice and immorality at Steveston. The constable endeavoured to do his duty, but what was one constable for a population of perhaps four or five thousand—Indians, whites, Chinese, Japanese, and negroes—and it was very discouraging to him to have case after case dismissed through some technicality. (J. W. Galloway, *The Missionary Outlook,* 1894)

The next year a New Westminster reporter, getting off the tram in Steveston, was immediately accosted by a man shouting "blackjack, blackjack." He reported: "It hardly seemed real that a gambling joint in a Canadian town would be allowed to keep a business rustler on the street, a condition of affairs that, according to report, puts even Seattle's palmy days in the shade." This reporter found eight gambling operations carrying on business with no attempt at secrecy other than a curtain at the doorway. He placed the blame for this state of affairs on the municipal government rather than the local police, who were frustrated by the lenient punishments handed out. Those who did not show up in court had their cases dismissed with bail forfeited, while those who did appear were fined and let go. For the price of a small fine, professional gamblers from Vancouver, Victoria, New Westminster and Seattle were able to make hundreds of dollars from the fishermen. Continued complaints to the municipal council finally resulted in a series of raids on the gambling houses, but nothing had much impact until Vancouver became a more attractive spot for the professionals. In one form or another, however, illegal gambling continued in Steveston into the 1960s.

By 1907 there was an attempt to shut down both gambling and prostitution in Steveston's Chinatown.

Now the trouble is that the Chief of Police has verbally notified the owners of the building that the Council ordered that there shall be no sporting women or gambling allowed on any of the premises and that the first sight of sporting women occupying the premises shall be arrested and persecuted to the fullest penalty of the law. (David C. Lew, *Letterbook*)

Prior to this date "sporting houses" were tolerated.

...what the owner wants is if you could have some arrangement with the Chief so to only fine the women during the whole canning season...same practice as years before. (David Lew)

It appears a solution was found to allow the houses to continue:

He says it is alright for you to go ahead and rent one cabin for one single woman in each row of the building but be careful not to sell any liquor, but they could give them away if people ask for them...(David Lew)

The influx of seasonal workers created a need for another kind of job, that of law enforcer. Richmond's first police chief, Herbert Drummond, was assigned to Steveston in 1891, and a jail was built. It took a while to establish a force equal to the task of keeping the peace during fishing season. For many years Steveston had only one constable to deal with all its problems. He was aided by the efforts of missionaries who followed the Native people to Steveston, but there was not nearly enough manpower to deal with Steveston's summers.

Then followed on our camp-meeting in June, when good was done and some souls were saved. After that came the salmon-fishing season. I spent three Sabbaths at the mouth of the Fraser River, two of which were while the strike was on. We had not a large number of Indians. The crowd came in the third week, when they began to fish, and the fish came in shoals about the last of July—from five to seven hundred in a boat. A fleet of about 3,000

fishing boats soon brought in more fish than the canneries could handle, and thousands were thrown overboard. At Steveston there was a great field from missionary work among the thousands of Chinese and Japanese, and I think more of the agents of that work should have been on the spot. Brother Tate, with several other Indian missionaries, took up the work when Brother Nicholas and I left. (Rev. T. Crosby, *The Missionary Outlook,* 1901)

A weekend night saw frequent fights and stabbings, keeping police and local medical staff busy.

The fishing season, on the Fraser River, B.C., has recently closed for this year, and has been almost a total failure; but our missionaries among those poor fishermen have found work in abundance, and it may be that, in their disappointment, they have been more ready to listen to the Gospel message. Dr. Large writes that, during the six or seven weeks of his work in Steveston, his office calls and visits totalled 850. His efforts were so appreciated that the Japanese Consul and the President of the Fishermen's Association both expressed their thanks for the faithful service rendered. (*The Missionary Outlook,* 1898)

Thefts, especially from the canneries, were very common.

Bold robbery—from a cannery—Steveston is excited over a serious robbery that took place there at midnight. Thieves entered Malcolm and Windsor's cannery and loaded their boats with $1,200 worth of new nets and made off with them. Malcolm and Windsor offer a reward of $150 for the apprehension of the thieves and recovery of the property. The steamer Delta, which had just cleared with lumber was chartered by Mr. Windsor and with police constable Alex Main on board went down towards Point Roberts in order to intercept the thieves should they attempt to pass that way. (*Victoria Daily Colonist,* April 1889)

In 1895 the police asked for a new jail because too many people had keys to the old one. The municipality was divided into police districts, and in 1897 Alex Main became constable for Steveston. In 1900 all police records were seized and examined on the suspicion that the constables were taking payments from Chinese merchants who were selling liquor illegally.

The municipality passed a public morals bylaw in response to the lengthy police reports of gambling, drunken behaviour and assault. Police chiefs were fired and/or quit in frustration over their inability to shut down the gambling and prostitution thriving in Steveston. The Sunday Observance law was drafted in 1891 and eventually passed in 1897, at the insistence of the Women's Christian Temperance Union and with the support of the local Presbyterian minister. The bylaw made it illegal to work, drink, sell merchandise, hunt or take part in almost any public activity on the Sabbath. Under the Public Morals bylaw passed in 1902, no one was allowed to use profane, obscene, blasphemous or grossly insulting language in the Township of Richmond. It was also unlawful to scream or create an offensive noise in public, and no vagrants were permitted to remain in any public place. The police regularly met incoming trams and sent anyone who was drunk or disorderly back to Vancouver. All the laws, bylaws and police resources of the little farming community were no match for the summertime onslaught of fishery workers, however, most of whom were single young men, eager to earn a relatively large amount of money in a short season and equally eager to have a good time. The chief of police, in his report for the year 1906, made of point of placing the blame for the town's crime on the seasonal workers.

The police record for 1906 also shows that although many convictions appear the offenses were largely committed by the transient population who arrive here in great numbers during the fishing season many of them bringing their bad habits with them yet including all of this we find our prevailing vices are drunkness [sic], gambling and vagrancy but to the grosser crimes named in the Criminal Code we can for A.D. 1906 conscientiously plead not guilty notwithstanding the slanders heaped upon this community by a few irresponsible hypocrites and mischief makers. From a study of outside police records compared with our own we find that the resident population of Richmond municipality is just as morally clean as any to be found in the civilized world. (M. Morrison, *Report of the Police Department,* Richmond Municipality, 1906)

Steveston's police even had the occasional murder to investigate. In the most spectacular case, the police chief himself was the victim. On Saturday, April 14, 1900, Alexander Main, the chief (and only member) of Steveston's police force, left home about eight in the morning to investigate the theft of a farmer's tools and overalls. When he hadn't returned by the next morning, his wife started an investigation of her own and found he had last been seen about ten the previous morning talking to a Chinese man. A search was organized by three Steves brothers; by Monday the entire population of the township was looking for the missing policeman with no success. Main's disappearance was reported to the Vancouver police chief, and a Vancouver detective named Thomas Wylie, a friend of the Mains, requested permission to search for him. Enlisting the aid of Lee Koy, president of Vancouver's Amalgamated Chinese Benevolent Societies and a major China contractor for the Steveston canneries, Wylie searched Steveston's Chinatown, eventually finding the man to whom Main had been seen talking. This man reported Main had been asking for another man named Yip Luck.

After acquiring four more assistants and a bloodhound, Wylie went to Yip Luck's cabin and found him there with another man named Chung Chee-Chung. Both denied seeing Main, but their behaviour aroused Wylie's suspicions to the point that he ordered the property searched. A steel-bladed brush hook with brown stains was found and the bloodhound, able to get a scent from the handle, led the searchers to a spot where the ground had been disturbed recently. Wylie and his men dug into the area and found the bodies of Main and his dog. Wylie took both suspects to the Steveston jail, then transferred them to the New Westminster Provincial jail to save them from being lynched by Steveston residents. In New Westminster, Chung Chee-Chung, who died a few weeks later from tuberculosis, confessed and implicated a third man named Ah Wong who was apprehended on his way to the US border.

Investigators discovered that Main had arrived at the cabin at midday while Yip was out. He began questioning the other two men about the tools and had found the missing overalls, when Yip returned and struck him on the head with the brush hook, killing him instantly. The three men killed, cooked and ate a chicken in the presence of the corpse to ward off evil spirits, and then Yip ordered the other two to carry the body outside for burial. When Ah Wong tried to get out of this job, he was

The Sockeye Hotel, 1910. Now called the Steveston Hotel, it was the only early hotel to survive fire and flood. (CRA 1977-19-25)

Steveston residents watching an opening, 1913. (F.D. Todd, BCARS)

threatened with the brush hook and went along with Yip's plan. They killed the dog and buried both in a hole dug outside the dyke.

Ah Wong acted as crown witness against Yip Luck and was found not guilty at his own trial. Yip's trial was the next day. Although he had no counsel, his guilty plea was accepted and he was sentenced to hang November 16. His last request was that the wooden board over his cell door giving his name, age and crime in both English and Chinese be used as a grave marker. In the presence of forty witnesses he was escorted to the gallows while a Chinese Christian missionary offered prayers for him. When the executioner pulled the lever, nothing happened. After a second attempt failed, a sheriff moved in to consult with the executioner about the problem. As the sheriff fiddled with the apparatus, the lever came unjammed and Steveston's most infamous criminal dropped to his death.

A number of churches had been established in Steveston, both to serve the permanent population and to offer spiritual resources to the transient.

The position and conditions of this place are somewhat peculiar, it is the centre of a farming district, and also the focus of the great salmon canning industry. The settled population in my district is about 400, but in the fishing season it amounts to from 6,000 to 7,000 of all nationalities. The farmers, who are a respectable class, are generally a poor struggling people, with but little to spare from their own needs for Church purposes, whilst amongst those connected with the fishing and canning interest there are few who care much for religion.

But whilst these people either cannot or will not do much to benefit the Church, it is the duty of the Church to do all she can to benefit them, and to this end it is quite necessary to have a Church, and to make it as attractive and pleasant as possible...(J.M. Donaldson, *Work for the Far West,* 1898)

The first church services in Richmond were conducted by a Presbyterian missionary from Ireland in a home on Sea Island. The island was also the site of a church built in 1886, whose congregation reached it by rowboat. Other Presbyterians assembled themselves at the Methodist Church at London's Landing, where a preaching station, tended by the minister at Sea Island, eventually grew into another Presbyterian church in 1902. These South Arm Presbyterians built a church at Steveston Highway and No. 3 Road. Several other Steveston families were Methodists and James Wood began a ministry to this congregation in 1885, building a church two years later at London's Landing. The Anglicans held services in the Steveston Opera House until their church, St. Anne's, was built in 1891 at the corner of Richmond Street and 5th Avenue. It was built on the foundations of a Baptist church started by Carvell Steeves, who lacked the funds to finish.

> At the mouth of the Fraser river is an island known as Lulu Island, on which live many poor and ignorant settlers, who are occupied in tinning or "canning" the salmon, which are so abundant in the river that they cover the bed, and force each other out on to the bank.
>
> The Rev. J.M. Donaldson has there built a Church, the dedication of which was the last act of the Bishop before leaving for England. And now a second and smaller Church, where the population is very thick, is imperatively needed.
>
> The sum of only £80 is required, and if half that sum be sent from England (Mr. Donaldson says), the people can be encouraged to give the other half themselves. Wood is cheap, and much labour is given free.
>
> The St. Anne's Society, Kemerton, a small parish Guild, consisting chiefly of the working class, took up this bit of work as their own, and are deeply touched by the kindness of Mr. Donaldson in resolving to call the Church "St. Anne's." (*Work for the Far West,* 1898)

The Opera House was also used by the Methodists and Baptists.

Homes and other buildings also served as temporary quarters for worship services.

> Before 1888 the ministers came over from Richmond Methodist Church and the Presbyterian Church on Sea Island. Mr. Wood from the Richmond Church used to preach in the afternoon, sometimes in the evening, in the Opera House. When the opera house was too cold, church was held in a small building on 4th Avenue. (Trites moved into this building afterwards).
>
> There was a Baptist Church in Ladner, and the minister used to come over about once every two months. He boarded at Steves when he stayed overnight. He used to have some services in homes like Mackenzie's at the corner of 4th and Chatham. This house was once used for an ice cream parlour owned by Maggie Quinn, and was later moved to 6th near Chatham. The Baptist minister was Mr. Williamson. Mr. Gay fetched the minister, and he preached at South Arm and Sunday morning in Steveston. After the Methodist Church was built in Steveston he asked permission to preach in the church, and Mr. Miller said he could if he didn't say anything against Methodist doctrine.
>
> In 1886 and 1887 services were held in homes. Mrs. Pearson like to have the ministers come to her home. Mr. Green preached at Mrs. Pearson's place, and the Baptist minister from Ladner also preached there...The South Arm Methodist preacher also held services in homes at South Arm. Mr. Green maybe stayed three or four years. He saw the need for a church at Steveston. The Port Simpson Indians wanted a church. The first meetings were held outdoors, or any convenient place....
>
> A.E. Green preached in the building that was an ice cream parlour. The church was so full that men were standing up at the back on seats because they could not see, also standing in the aisle. It used to shake in a heavy wind and the fire smoked in a northwest wind...The Ladies Aide put a big homemade cumbersome pulpit in the church once it was built. It may be the one taken to the Lodge meeting room in the Orange Hall. John Tilton put the carpet on the platform at front of the church and put the railing around it...After the church was built in 1894, A.E. Green was the first minister. Two hundred dollars was still not paid for, and they collected it from the canneries. (Ida Steves)

The missionaries and ministers serving early Richmond usually served other small communities as well, and rotated visits with clergy of various denominations. A small church built in 1870 on the mainland functioned as a union church for Anglicans, Presbyterians and Methodists.

COMMERCIAL HOTEL.
COMMERCIAL-RESTAURANT
MEALS ALL HOURS

Downtown Steveston, early 1900s. Note the boardwalks. (BCARS 82292)

Shortly after Bishop Sillitoe's arrival in this Province, about 1880, he came down...to hold services in this small Union Church and at the first of these services a rather unpleasant occurrence took place. The congregation was made up largely of Methodists and Presbyterians for there were but few Anglicans in the settlement, and, as was their custom, no matter what denomination the clergyman belonged to, each sect took the postures, during prayer, to which they were accustomed in their own churches. These postures, especially that of the Methodists', who knelt with their backs to the pulpit, displeased his lordship very much, who diverged from his service to express his displeasure, which was done in a very caustic manner, and as some expressed it, "not in a Christian spirit." One of the leading Methodists of the settlement rose and said that most of the people present came to hear the Gospel of Christ preached and hoped they would not be disappointed. The few Anglicans who were there, not being high Churchmen, wrote a letter of protest to the Bishop, which did not temper his severity and to which he referred at his next visit as being an effort to charge him with Papistry.

At his following visits to this settlement, which were few, his hearers were less than at his first service, for the Presbyterians were the descendants of forefathers who had withstood the onslaught of the sword-proud Royalist Claverhouse, or of those across the Irish Sea, who could neither be persuaded by the eloquence or subdued by the persecution of Bishop Jeremy Taylor. Nor did his severe remarks lead the Methodists to modify those simple forms of worship which their forefathers had adopted instead of the rituals of the Anglican Church. (Thomas Kidd)

The Salvation Army sent representatives to Steveston during the fishing season and had a parade down Moncton Street, past the bars and gaming halls, every Saturday night. In 1925, when the Canadian Methodist, Presbyterian and Congregational churches consolidated to form the United Church, some Presbyterians elected to retain their former denominational affiliation and attend the Sea Island Presbyterian Church. When that

The Salvation Army band on its Saturday parade through downtown Steveston. The Vancouver General Store, pictured here, was established in 1899 by Simon Peretsky. It sold groceries, tobacco, drugs, sundries and "misfit clothing." (CVA)

burned, they chose not to rebuild; some of the congregation joined Marpole Presbyterian and the rest Richmond United.

In the early 1890s a Catholic mission was founded on the property of the English family, the first canner in Steveston.

> The Catholic church was built on the English property, which ran up to where the extension of Chatham Street is now. Ida went to the church opening. The bishop consecrated it. They sat down in the church a little early. Then the bishop was robed for the service, and everyone had to leave. The bishop sprinkled Holy Water on the Church, on the floor, the walls, all the pews. The service was all in Latin. They avoided the Presbyterians.
>
> Mrs. English had been sent to a Catholic convent and became a Catholic. Mr. English was not a Catholic, but when he became sick, he said, "Make me a Catholic quick." The church really was built in time for fishing which always started the first of July. It was fifty-nine years old in 1963. It was started with money from England. (Ida Steves)

In 1932 two nuns from New York were sent west as missionaries. Services were held on the upper floor of a house on 2nd Avenue, the lower storey serving as a convent, kindergarten and Bible study area. In 1933 a rectory was built on 2nd Avenue and in 1949 St. Joseph's was established as a parish.

In 1928 a Buddhist temple for the Japanese population was built south of Moncton on First Avenue. Earlier plans had been thwarted by the white population. Christian missions among the Japanese had made a few converts, most of whom affiliated with the Methodists and later the United Church.

During the fishing season several missionaries accompanied the Natives who came to work in the canneries. Their spiritual services were appreciated by the permanent residents as well.

> The missionaries came down with the Indians for the fishing season. There might be several here at the same time. Before the church was built, Mr. Tait was here and he was staying at the Richmond Hotel (built in 1890) and there were only newspapers on the windows. He insisted on having real curtains and they had to get real curtains and put them on the windows. The Nicholsons were missionaries who came from Vancouver Island.
>
> After they built the church and the parsonage, the missionaries would stay in the parsonage. The white people would go in the back door and sit at the front of the church, as close to the back door as possible. The church would be packed with Indians. Mission services were held in the morning, and other ministers came in the afternoon.
>
> Dr. Large would be here during the summer season. We always tried to hear Dr. Crosby because he was such a good preacher.
>
> The missionaries also held meetings at the Indian camps at each cannery.
>
> At one time there was a young lady missionary. She preached near the Gulf of Georgia cannery (near 6th and Chatham). George Smith was quite taken up with her. We went down a few times to hear her, but it was the middle of summer with hired men to be fed, and it was very hard to get out to services in the morning. The missionary's mother was with her. This was before the church was built.
>
> Another time a woman came to Steveston and started revival meetings for the Methodists...They must have held meetings every evening for six weeks in the Opera House. She decided Steveston must have a Methodist Sunday School. (Ida Steves)

Women's groups associated with the churches played a major role in the social life of Steveston. The women got together for various mission projects and, more importantly for the life of Steveston, to organize children's activities and schools.

> The Methodist and Baptist Sunday School lasted no more than a few years. There were very few children except during the fishing season...The adult bible class was more successful, but finally they decided there were not enough people to keep one denomination going, and started the Union Sunday School in 1895. The Methodists also had the Epworth League...(Ida Steves)

Steveston children were originally taught at home or at classes held in the Opera House. By 1887 there were enough families settled on the south side of Lulu Island to create a demand for a school. The Minister of Education approved a petition to create a school district, trustees were elected and a building was planned. This school first held classes in the Methodist Church on London's Landing, and moved a year later to a school building at the corner of No. 2 Road and Steveston Highway. As the population grew during the 1890s, a new Steveston district was created, and a new school was built in 1897 at the corner of Georgia Street and 2nd Avenue.

> **About the latter part of 1885 Virgie English began to teach her younger brothers in her home...Soon other children in the area were attending also...About the first part of 1887 the English family went to California, and Mr. Robinson, a teacher from Ontario, came to Westham Island, and he taught at the new church at London's.**
>
> **In the fall of 1888 the one-room schoolhouse was built at the south east corner of No. 2 Road and No. 9 Road. One of the first teachers was Mr. Muir. Mr. Jean Baptiste Hebert, a French Canadian who the boys called John the Baptist, taught there while he and his family lived on the Gerrard farm.**
>
> **McKinney had 160 acres along No. 2 Road and 160 acres on the north side of No. 9 Road. He kept a big bull near the fence and the children would have to get into the ditch, which had no water, and walk past the bull. The school was just a little brown schoolhouse...This schoolhouse was later moved from its site to Pleasant Street in Steveston for a Presbyterian Church.**
>
> **The first classes [in Steveston] were held in the Opera House...Later they used Maggie Quinn's ice cream parlour on 2nd Avenue...The first one-room school in Steveston was built around 1897, and this was replaced with a two-room school in 1905, the one-room school being used for storage and a playroom. These schools were on 2nd Avenue and Georgia Street, in the same spot where Lord Byng stands now...**
>
> **In those days there was seven years public school. In the first year children learned the two primers, then one year for the first reader, and one year for the second reader. All these children would be in one room in a two-room school. The third reader took two years, and a student was still doing the fourth reader when he wrote entrance exams for high school. Students started both in September and in January, depending on their birthdays. In high school you had to pass every class to go on to the next year.**
>
> **When the Inspector came, he asked arithmetic questions. Once when he was making one up, he asked how much potatoes were a ton, and George Fentiman said $10 a ton to make the question easier. (Ida Steeves)**

Teachers lived in boarding houses or roomed with families. School furniture consisted of desks, chairs and a wood- or coal-burning stove, and there was an outhouse. Each child received a slate, a piece of chalk, a bottle of water and a rag. They had to buy their own textbooks. In 1906 the four school districts in the municipality were consolidated into one, and more schools were built. Children could complete Grade Seven in Richmond; to continue their educations they had to go to Vancouver or Ladner.

> **In 1909...students went over to Ladner to write the entrance exams...School had been closed for two months with scarlet fever, and Mrs. Steeves told Mr. Davidson she did not think it was any use the children writing the final exams because they had missed so much time at school...The students from Steveston and English school went over to Ladner by boat...they were there three days...Samuel McElvanie was the principal then and...he helped them write the exams again at Christmas. Three of them...went into King Edward and passed the entrance exams...The exams at King Edward in Vancouver were not the same as the exams they wrote at Ladner. The Ladner exams came from McGill University. If the Richmond students had had the same exams as the Vancouver schools, they thought they would have passed. (Ida Steeves)**

This situation was made easier when, in 1908, the two upper rooms in Bridgeport School were designated a high or "superior" school.

> **Bridgeport was used as a high school for all of Richmond. When the electric tram went through in 1905 students would take it to the school. Tickets were 1/2¢ each...High school was the equivalent of grade eight, nine, ten and there were less than a dozen students all together. (Ida Steeves)**

Enrolment at the Steveston School, later

renamed Lord Byng after Canada's Governor General, grew steadily over the years. The Japanese community, who were forced to operate their own school on No. 1 Road, wanted to add another four-room school in the 1920s so their children would have the same opportunities as the white children. They offered to make a major contribution towards building the new school if they could also use it for an after-hours Japanese language school. The original, pioneer Steveston School had been fully integrated with Chinese, Japanese, Native and white children attending together. But after a large population increase of Japanese families between 1909 and 1925, nearly all the Japanese children in Steveston went to school at the new Japanese school on No. 1 Road, a Japanese Buddhist hall on Chatham Street or the parish hall in St. Anne's Anglican Church, with teachers provided by the school board. The school board ruled that only "known residents" could attend Steveston school and this excluded the children of Japanese cannery workers, many of whom lived on cannery property outside the dyke and thus did not pay property taxes. Special assessments were levied on them to pay for social services such as schools. By the time the school district built a new, fourteen-room school on the Steveston site in 1930, again with the financial help of the Japanese population, all the Japanese children were attending either the new Lord Byng or one of the other general schools, as well as their Japanese language schools.

As the community grew, more public health concerns emerged, especially relating to living conditions in the nonwhite communities. In 1891 the municipal council appointed a health committee to document diseases or epidemics. In 1896 a permanent public health officer was appointed and the following year he documented

The Japanese school at Steveston.

thirty-eight cases of typhoid among twenty-six Japanese and twelve Natives. In 1902, two cases of diphtheria in Steveston led to an investigation of the cannery shacks. In 1909 there were two cases each of scarlet fever and typhoid in the Steveston Japanese community; a 1910 outbreak of typhoid was traced to diseased salmon found in one of the ditches. There were also cases of infantile paralysis in Japanese children. Typhoid appeared again in 1911, and in 1917 two immigrants from Japan were turned back on the suspicion they had tuberculosis. An outbreak of measles occurred in 1918 and the threat of the Spanish flu in 1918–19 led the provincial Board of Health to close all schools, churches, theatres and other public institutions. Richmond had seventy reported cases of the deadly virus but few casualties. The racetrack clubhouse served as a temporary infirmary. In 1920 spinal meningitis was found in a Chinese child, 1926 saw an outbreak of diphtheria and mumps, and in 1928 the health officer reported three cases of smallpox. Contagious diseases were of concern to the local medical community, but in practice they were kept much busier with farm and cannery accidents, and the stabbings and other violence during the fishing season. Drownings were very common, especially during the summer months.

An American dentist, U. Yamamura, visited Steveston and was upset by the working conditions he saw at the Phoenix Cannery. He pressed for the building of a medical facility for the Japanese community. The result was the Japanese Fishermen's Hospital, constructed as a chapel, school and eighteen-bed hospital in 1896, financed and administered by the Dantai (the Japanese Fishermen's Benevolent Society). Located on No. 1 Road near Chatham, the hospital cared for Japanese and white patients and charged a membership fee of eight dollars per family. Apart from this facility, if patients needed hospitalization the doctor had to drive them to the Marpole Infirmary or St. Paul's Hospital in Vancouver. In September 1942, after the Japanese were evacuated from Steveston, the Fishermen's Hospital was renamed Steveston Hospital, the only local health facility until Richmond General Hospital opened in 1966. When people died, funeral services were held locally, but interment was in Vancouver because Lulu Island's high water table made cemeteries impossible in Steveston.

The Japanese hospital, 1908. (VPL 2173)

Although farm life in Steveston's early days occupied most of a family's time and energy, there were many recreational moments too. Visiting family and friends was a traditional pastime with its own schedules and rituals, no doubt growing out of pioneer loneliness.

Mrs. Manoah Steves was a woman of the most placid and hopeful mentality, which helped to sustain her through all the privations and inconveniences of pioneer life and in the more serious trouble that come to her in the death of her oldest daughter, Josephine, at the age of 26, of her oldest son, Wm. Herbert, at the age of 39, and her youngest son, Walter, at the age of 36. Mrs. Steves was of a most retiring nature, but most congenial with the few friends that the opportunities of pioneer settlement gave her. (Thomas Kidd)

Race day at the Brighouse track in Richmond. Going to the races was a popular amusement for Steveston residents. (CRA 1977-2-42)

In his history, Kidd singled out one woman in particular "whose visits to her neighbours helped those of her own sex forget their isolation and find compensation in that new settlement as compared with that of city life." There were also numerous church-sponsored teas, picnics—sometimes at Stanley Park or one of the Gulf Islands—and other social events. While the Opera House existed, it was the site of many dances, revues, professional entertainment and locally organized socials. Election days were also social occasions.

The municipal hall was at the corner of No. 3 Road and River Road...When there was an election, people went over there to vote. The polls closed at 4:00 p.m. so they had to hurry the horses to get there in time. It was quite a ride over corduroy roads. All the people who had voted stayed there to see how the voting came out. The room was so thick with tobacco smoke they could hardly see. (Ida Steeves)

Organized sports were very popular, especially lacrosse, which was introduced to Richmond players at a field near Eburne sawmills in 1907. By 1912 Richmond had two teams, the Fisheaters and the Muskrats. Rugby, football and baseball teams also looked for competition beyond Richmond. Horseracing was a very important activity. Minoru Park, built in 1909 and named after King Edward VII's horse which had won the Epsom Derby, attracted racing fans from all over the lower mainland to Richmond. The one-mile oval was a very popular facility because of its superior turf. Minoru's popularity led to the construction of another one-mile oval at Lansdowne in 1923. As provincial

The Steveston Tyees, a local lacrosse team, 1937. Back row, left to right: Mick Burdett, Alf Hepworth, John Spargo, Bud Young, R. Jack (assistant coach), Bert Hart, Jack Deagle, Harry Jarvis. Middle row: George Mackie, Bill Wallace, G.W. Skinner (manager), Bill Gilmore (coach), Ken Elston, Ben Rose. Front row: A. Vincent (trainer), Bob Francis, Kaoru Kobayashi, Toru Kobayashi. (GGCS 1977-7-6)

gambling restrictions decreed that no race track could have a season longer than two weeks, the racing circuit was at Brighouse for seven days, Lansdowne for seven days, Hastings Park for seven days, back to Brighouse for a week, followed by a week each at Lansdowne and Hastings. The final two weeks of the season were on Vancouver Island at Colwood and Willows. A special tram station serviced thousands of spectators who enjoyed a day at the races. Horses from the best American stables came to run. The Minoru track was closed during World War One, but reopened in 1920 as Brighouse Park. When the horses weren't racing, other events were held at Brighouse, including boxing matches and polo; it was also the site of May Day celebrations. This festive event, sponsored by the Richmond Agricultural Association, included a parade, the crowning of the May Queen, the maypole dance and other folk dances, and marching drills by the community's school children, who wore coloured paper hats and followed marching patterns that spelled out the words "Jubilee" and "Richmond" (the children then assembled themselves into a Union Jack design). Afterwards there were foot and buggy races on 2nd Avenue in Steveston and a community dance.

There would be a sports day, put on by the community, and the school together. In Steveston the races went from the school to the

church on Chatham and 2nd. This was the 24th of May. On the 23rd of May, which was Empire Day, everyone went to Bridgeport, and we sang the Maple Leaf, Our Emblem Dear; 3 Cheers for the Red, White and Blue; and God Save the King. We took our lunches and had a sports day. Later on we went to Minoru Park, had decorated floats, the young ones had maypole dances, and the older ones folk dances like "Gathering Peascods" and Japanese dances in beautiful kimonos borrowed from Vancouver. (Ida Steeves)

There was also the Japanese Kite Festival celebrating Boy's Day, held in the old Buddhist church. Richmond had a popular rifle range and hunting was a very common activity, both for sport and as a source of food. Band practice also occupied the leisure moments of many young Steveston residents.

There were no parks in Steveston at this time. People met and socialized in homes and halls, and in streets and shops. As the community became more family-oriented, great efforts were made to secure a piece of land for a playground, specifically the three acres near the BC Electric Railway station at the corner of Moncton and No. 1 Road. Finally, in 1923, the Steveston Athletic Club and World War One veterans set up fundraising booths, selling tickets for a "Queen" contest to be held among

Local lacrosse heroes. Left to right: Bill Gilmore, unknown, unknown, Mike Parker, unknown, Vinnie Hartney, Sam Gilmore (Terra Nova Sam), Johnny Blaire, unknown, Sam Gilmore (Steveston Sam), Bob Dixon, Eddie Gilmore, Bill Morphat, unknown, Wilf Tolley. (CRA 1987-104-48)

three local girls, which ended with a dance at the Opera House. Their efforts raised the bulk of the money needed and by July Steveston had its first park. The municipality supplied free water and regulations for the preservation of law and order. In the 1930s the Richmond Athletic Club built a lacrosse box with wood donated by the Imperial Cannery and a net found in a boathouse. House teams from the Imperial and Phoenix canneries practised there and became formidable opponents against teams throughout the province.

By the end of the nineteenth century Steveston was a prosperous community, well serviced within its own bounds and well connected to its neighbours. It was surrounded by thriving farms and profited from being the centre of the fish-canning industry. The farming population of Steveston, in contrast to the rest of Richmond, had evolved from a largely white undertaking to a mixture of European, Chinese and Japanese enterprises. There were about fifty farms, most averaging five to twenty acres. The largest farms were owned by the Steves (approximately three hundred acres), Hong Wo (two hundred acres), Mah Bing (one hundred acres), Lun Poy Brothers (fifty acres) and the McKinneys. Some of the larger farms leased parts of their properties to Japanese farmers. In the summer the field work was done by Japanese, Chinese and white labourers working side by side.

For the farming community, Steveston had become a comfortable settlement and a good place to raise a family. David Jelliffe combined the information in over one hundred interviews of Richmond pioneers to provide the following description of a Steveston childhood at the turn of the century.

> Some of the roads were made out of planks, some of gravel, and the rest little more than cow-trails—dirt paths, really—that turned into mud tracks during the rainy season. This was Richmond around 1900 when a man could raise a family by fishing or farming. What was it like to be a child in those days?
>
> ...children were generally born at home. The doctor would come to the house and with the help of a neighbour woman or sometimes a travelling nurse, make sure you were born and your mother was OK. Relatives or neighbours would help out until your mother was up and around again...There was also no electricity and running water so everyone used the outhouse in the back and coal oil lamps. Wood stoves provided heat and the means to cook. Later, when the coal barges unloaded in Steveston, a man with a wagon of coal would come by the house and unload so many sacks of coal. Wood was the main fuel though.
>
> That was the first job children had to do for the family: keep the wood-box in the kitchen full from the wood supply outside. As one grew older, the jobs or chores became a regular part of everyday life. Of course, the question "do I have to" was heard all over Richmond. Every father and almost every mother said "YES" because in those days there were no "rich kids". Later, the chores of pulling weeds in the family vegetable plot, feeding the chickens and pigs, and running errands were things the family depended upon you to do. If you lived near the Fraser River, many boys and girls would go with their fathers or a hired man to collect drift wood to be cut up, loaded on a wagon, and lugged back to the wood shed or wood pile next to the kitchen door. A year's supply of wood was often collected this way. Children usually enjoyed these outings because there was so much to see and do besides just gathering up wood along the River's edge; watching birds, picking flowers, throwing stuff as far as you could into the main channel, fooling around the edge of the water without getting your shoes wet, and climbing on top of some of the huge logs and stumps washed up on the shore.
>
> Children born on a farm had more than a few chickens, pigs, and an old cow to contend with. Many of the early Richmond farms had 50 or 60 milking cows, beef cattle, and up to 8 horses. Some horses were only used for plowing or working in the fields, others only for pulling a wagon (usually as a team of two), others for the buggy or democrat...boys soon learned how to milk a cow—starting with an old cow or one that was drying up—and when they got good at it, the boys were then allowed to milk what is called a "high producer". On some farms the women milked cows as well so the girls learned from their mothers. All the cows and horses had their own names and would sometimes pay attention to you when you talked to them. The big farms had hired help who usually lived on the place and ate with the family. These men could be white, or Hindus, or Chinese; many old-timers today

talk about how much they learned as children from these patient people. Learning how to ride bare-back, holding down the hay stacked high in a wagon, collecting eggs, driving the cattle to a new pasture, pouring the fresh warm milk into the cooler, learning how to skim cream and how to make butter in a rotary churn, learning how to shoe a horse; all these things were part of life on a Richmond farm. The food was fresh and good, the chores were tiresome but had an end, interesting things happened, but there was very little money to be spent in the general stores.

In those days, a ride on a wagon into Vancouver took just about all day...the trip would begin by going north along No. 1 Road to River Road. Then along by the dyke next to the Middle Arm, past the old Provincial Cannery (later Easterbrook's Flour Mill) at the end of Bridgeport Road, and over the Lulu Island Bridge. This bridge connected Lulu Island with Sea Island and swayed quite a bit. Then past Grauer's store...and the Eburne Post Office...on over the Marpole Bridge. From there it was all uphill along Granville to 41st Avenue. In those days, the road along Granville Street was through a pretty heavy forest. Huge trees on both sides of the road were being systematically cut down and the land cleared....

From 41st onwards, its all down-hill to the old Granville Street Bridge. Once in town, the wagon is driven to where you wanted to unload. If you're going shopping with your mother while your father is doing business, the main stores were Woodward's or Spencer's (Eaton's). The movie houses that developed

The Hepworth family outside Dr. Hepworth's home.
(CRA 1987-104-50)

later were the Pantages (Lyric) and the Orpheum. A small movie house was started in Steveston (the Steva) on the site of a Buddhist church, here silent movies were shown, but movies in Steveston didn't amount to much.

Children born in Steveston often didn't have a horse to ride but there were other things to do in their spare time. For example, at the end of No. 1 Road there was the government ferry wharf where the ferries from Vancouver Island use to tie up and unload wagons, horses and people. Children would, as well, walk along the dyke from No. 1 Road to London's Landing (at the foot of No. 2 Road) to see what was going on along the edge of the River. There were several small stores along the dyke where one could buy a penny's worth of candy. The most famous of these was called Hong Wo...and old-timers today recall the fantastic assortment of goods sold there. There was also a number of boat ramps to be explored and empty cannery houses one could prowl around if it wasn't the fishing season. Otherwise you could hike out to Garry Point where the old tree has a big lantern tied up near the top to mark the entrance of Steveston Harbour on one side and the big hole in the jetty on the other. This hole is called "the gap" for it allows fishing boats to float away from the Fraser River into Sturgeon Banks by means of a natural slough. During the fishing season there are a lot of salmon in this slough. This is a great place to swim, in "the gap", and many boys learned how to swim right here.

Cannery workers' housing in the Chinatown area. (CRA)

Boys were the only ones to swim out there off the jetty at Garry Point because hardly anyone had a bathing suit so you swam in your "birthday suit". But there were other places a person could swim in Richmond....

In winter there was usually skating on both the North and South Arms of the Fraser if the winter was cold enough. Some children have been known to skate all the way from Sea Island to New Westminster and back on the North Arm. Kids played hockey under the Fraser Street Bridge. If there was a heavy snow, the East Richmond farmers covered their wagons by replacing the wheels with sleigh-runners enabling the horses to pull the wagon on top of the snow. Many farmers are known to have brought wagon-loads of feed, lumber, and other materials to and from Richmond by crossing the frozen river. A great experience: travelling with your father over the ice, watching the clouds of breath from the horses, and listening to the deep cracking sound of the ice as the floes shifted and bent under the pressure of the ice up River. However, there was no skiing or sledding of any consequence for Richmond children because there were no hills in Richmond. You had to go to Vancouver or the North Shore if you wanted to sled.

Even today, the stretch of water between Steveston and Steveston sandbar (now called Shady Island) usually freezes in winter allowing children to skate or play hockey. The rest of the River doesn't freeze over solid anymore because river traffic keeps the ice from freezing solid.

Things were much different in Steveston during the fishing season. Hundreds and hundreds of Indians and Chinese came to fish or to work in the canneries. Children were usually kept home on Friday and Saturday nights because there were fights and brawls and an occasional street killing among the hard-drinking fishermen and cannerymen. You must remember that all fishing had legally stopped on Friday night at 6 p.m. and wouldn't start up again until Sunday night at 6 p.m. when the gun went off.

During the fishing season, at night, one could hear celebrations from the Indian encampment over by the dyke next to 7th Avenue in Steveston—throbbing drums that lasted all night long. Lying in bed, one wondered what was going on. Children were absolutely forbidden to go over to see the huge bonfires and the usually placid men wildly jumping about with the women in long dresses swaying in the background and even sleepy-eyed Indian children huddled in the shadows watching the old tribal rituals and games. Some Steveston children disobeyed and sneaked over to the Indian encampment to watch. Of course, being caught over there meant a licking but some felt it was worth it!

During the fishing season, some Steveston boys would go to the back of one of the canneries and dig about on one of the big scows that was getting filled with offal. Offal is the name for all the parts of the fish that aren't canned. Thus, guts and tails and heads are tossed into these scows. The boys would search out the fresh salmon heads and with a sharp pocket-knife cut out the cheeks of the salmon. Six heads yielded a dozen cheeks and you brought them home to make a good supper for the family. Your Dad might say "salmon again?" and glare at your mother but when she tells him you brought it home, he'd look over and say, "well, I like salmon; where'd you get it, son?" and you would tell him.

The scows, when filled, were towed over to Williamson Island (behind Kirkland Island in the South Arm across the way from Steveston), where there was a fish-reduction plant. Around World War I the plant burned down and until the individual canneries started their own reduction process (to get fish oil and meal) the offal was towed out into the Strait of Georgia and dumped. So, the scows smelled; that didn't stop the boys though....

Living in Steveston at the turn of the century meant living in a B.C. town that was in many ways bigger than Vancouver. The 3 and 4 masted schooners from around the world docked in Steveston to load salmon and lumber. A big ship would come in and all the kids would troop down to look at the strange sailors—many tattooed—who seem to speak about 5 languages at once. Some guy on board ship would wave at you to "come aboard" and you'd all troop up the rickety gangplank to look at caged parrots or uncaged monkeys who chattered at you and jumped about on the end of their chain. B.C.'s first zoo came and went with these ships and only later did Stanley Park get the buffaloes, bears, and penguins.

Early farming fashion in Steveston. (CRA)

You must understand that sailing around the world under "canvas" was a really rough way of life and it has been said that children at this time have known sailors who've "jumped ship" and hidden out with the farmers in Steveston or Ladner; when the first mate from the ship with a gang of tough-looking sailors came up to your place, looking for a deserter, your father or uncle would say quietly to you but in a voice as hard as nails, "don't tell," you wouldn't. You were more scared for that poor man hiding in the barn than you were even of your own father.

The sailors were never found and usually worked for their keep on the farm that hid them and in a while they became good citizens of Richmond and your friend for life....

In those days, your father didn't have much time for you because he wanted to keep the family alive and had to work 12–15 hours a day. If you ever came home from school bawling that the teacher had strapped you, there was a good chance your father would strap you again; thus one learned not to talk about school, at least about that part.

...kids caught muskrats too and cut off the tails which they brought to the Municipal Clerk and got 10 cents for each tail. The reason there was a "bounty" on the muskrats was that the animals made their little hovels in the dykes, and this weakened the dyke at that point. Flooding due to a broken dyke was often caused by the muskrats, thus the 10 cent bounty....

Trapping muskrats wasn't the only way to catch wild animals that were all over Lulu Island. There were rabbits, foxes, deer, an occasional bear, and game birds. Fathers, big brothers, or neighbour boys would show you how to handle a shotgun and as the really big flights of ducks or geese crossed over your place, you learned how to get a couple of birds for supper. Hunting pheasants was another matter, particularly in the fields of a large farm, because you had to be alert and ready when a covey broke cover.

Before we get all steamed up about shooting birds you should understand that Richmond people in the old days had to depend entirely upon themselves to stay alive. Thus farming, fishing, gathering wild berries, and hunting were the things to do to get food. As mentioned earlier, there was very little actual money circulating about so one would trade potatoes, butter or eggs for meat at Grauer's store....

Living in Richmond around 1900–1920 wasn't easy, but at the same time it wasn't terrible. There were the May 24th races, picnics to English Bay, Strawberry Festivals, the New Westminster Fair, and finally, a good chance to find money if you crawled under the plank sidewalks in Steveston in front of the 5 Hotels on a Sunday morning. All in all, living in Richmond in the old days seemed like an exciting time to be a child.

Child standing outside the Royal Bank in Steveston. This building now houses the Steveston Historical Society. (CRA)

Chapter Three:
The Fishing Community: The Pioneering Years

The familiar contents of Grocers' shops are sent from this remote corner of the world. After various processes of washing, cutting, boiling & roasting, as regards the salmon, & rolling, clipping, soldering & pasting, as regards the tin, both are ready for the far away multitudes of English work-people.

"We eat what we can & can what we can't," is a much hackneyed saying. But a certain Scotchman, it is said, carried the saying home & repeated it one day in his narrative of the big things he had seen. "They have a saying out there over which they laugh greatly, but I dinna see ought to laugh at; they say: "we eat what we can, & what we can't we tin." (Rev. Appleyard, *"Missionary Work at Port Essington"*)

The salmon fishermen on both sides of the line are of many nationalities, most maritime nations of Europe being represented and also the Japanese. A large proportion of Indians and half-breeds, and some negroes are also employed. The Chinese, however, while they compose the bulk of the help in the canneries, have participated only to a very slight extent in the fishing and not at all in Canadian waters. (US Commission of Fish and Fisheries, *Report of the Commissioner for the Year Ending 1877–8*)

Japanese women with babies on their backs, filling cans at the Imperial Cannery, 1913. (F.D. Todd, BCARS)

The first Steveston farms were established about the same time fish canning enterprises were developing upriver. The canners soon moved their establishments to the Fraser's mouth, adding a very prosperous industry to Steveston's economy and fuelling W.H. Steves's hopes for the future. For Stevestonites, the smell of fish was and is the smell of money.

Five species of salmon spawn in the Fraser River and its tributaries. But the Fraser River canners initially depended on the sockeye, and to a lesser extent the coho, as the mainstay of their packing operations because the nineteenth-century British market accepted only red fish. This preference was due in large part to the fact that the first pink salmon shipped to England were badly packed, lacking the firmness and oil content by which the British consumer had learned to judge the quality of salmon.

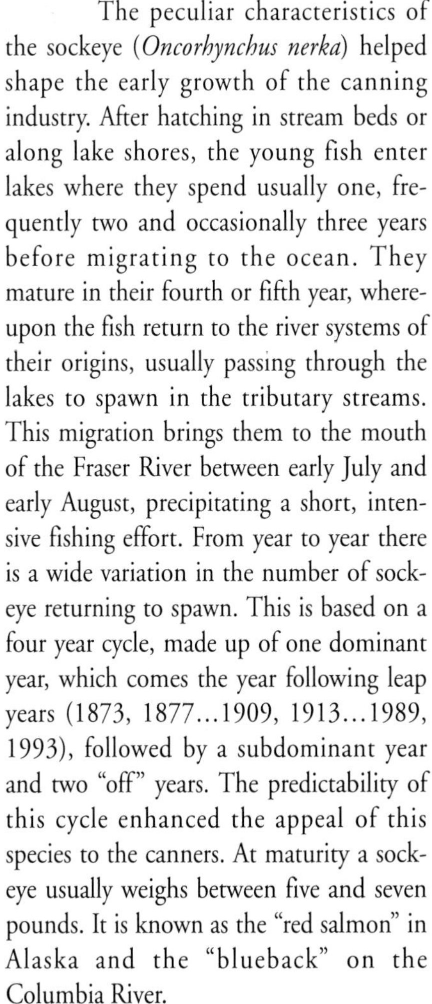

The peculiar characteristics of the sockeye (*Oncorhynchus nerka*) helped shape the early growth of the canning industry. After hatching in stream beds or along lake shores, the young fish enter lakes where they spend usually one, frequently two and occasionally three years before migrating to the ocean. They mature in their fourth or fifth year, whereupon the fish return to the river systems of their origins, usually passing through the lakes to spawn in the tributary streams. This migration brings them to the mouth of the Fraser River between early July and early August, precipitating a short, intensive fishing effort. From year to year there is a wide variation in the number of sockeye returning to spawn. This is based on a four year cycle, made up of one dominant year, which comes the year following leap years (1873, 1877...1909, 1913...1989, 1993), followed by a subdominant year and two "off" years. The predictability of this cycle enhanced the appeal of this species to the canners. At maturity a sockeye usually weighs between five and seven pounds. It is known as the "red salmon" in Alaska and the "blueback" on the Columbia River.

Chum salmon (*Oncorhynchus keta*) were long regarded as a nuisance by fishermen, although the Natives always used them to some extent. They are the last of the Pacific salmon to appear during the fall in spawning schools. They usually weigh eight to eighteen pounds at maturity, in their third or fourth year of life. In 1897 the Japanese started dry-salting chum salmon on the Fraser for the Japanese market and for use in the Yukon as dog food. Possibly for this reason, or perhaps because the breeding males develop very large teeth, this species is sometimes called "dog" salmon. In 1900 the sockeye run was very small and a good price was offered for lower grades of salmon; 105,000 cases of chum were canned.

The *Oncorhynchus gorbuscha*, commonly known as the pink, were long considered inferior in value for canning because of their light-coloured, soft flesh. However, as the sockeye became scarcer and demand for cheaper grades of salmon increased, pinks eventually became important and today nearly the entire catch is canned. They spawn from late September to early November, when the males develop large humps on their backs, which led to their being called "humpbacks" in the early days of the fishery. The young go to sea as fry and reach maturity at two years, when they generally weigh between three and five pounds.

The young of the *Oncorhynchus kisutch*, the coho, remain in streams for a year before migrating to the ocean where they grow rapidly, usually reaching maturity at the end of the third summer, weighing from six to twelve pounds. In the Strait of Georgia young coho at the beginning of their third year have deep blue backs and bright red flesh and are also known as "bluebacks". In the US they are usually called "silver salmon." Some of the catch is canned, but most is marketed fresh or frozen. A US commission at the turn of the century noted: "The Indians prefer this species to the sockeye for their own use, probably because it is more readily cured by their process of drying." The report also noted that coho were not often canned unless there was a shortage of sockeye, but they were extensively salted on the Fraser for export.

The Chinook salmon (*Oncorhynchus tshawytscha*) begins its spawning run in late spring, reaching maximum numbers in rivers by early autumn. They take four or five years to mature, at which time they may weigh anywhere from ten to fifty pounds. The catch is canned, frozen, smoked and dry-salted, and large quantities are sold as fresh spring salmon. They are popular with sports fishermen in the central coast area where they are called "tyee"; in the US they also go by the names "king"

Sockeye
Oncorhynchus nerka

Chum
Oncorhynchus keta

Pink
Oncorhynchus gorbuscha

Coho
Oncorhynchus kisutch

Chinook
Oncorhynchus tshawytscha

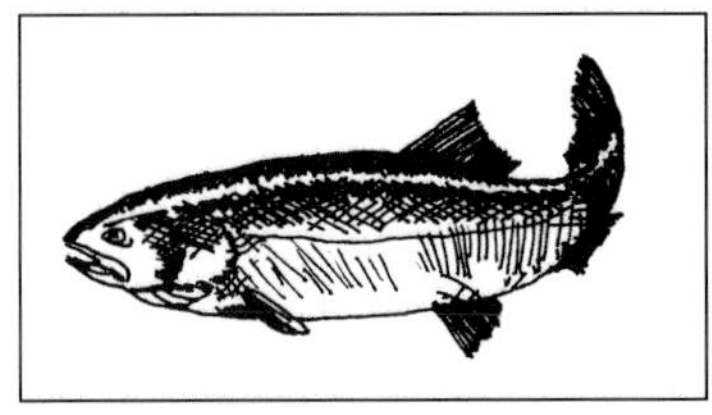

(Lionel Trudel illustrations)

Fishermen and boat-pullers setting their gill-nets on the crowded fishing grounds at Garry Point, c. 1900. (DS)

and "quinnat" salmon. In Canada they are popularly known as "springs." This was the species most favoured in the aboriginal fishery because its low fat content made it easier to preserve by air drying and smoking than the oilier sockeye.

Before the development of fast and efficient transportation systems, BC was too distant from the large consuming centres to do much more with its salmon resource than to catch it for local consumption. Gradually, methods of preserving the fish developed: first salting, then canning, freezing and the shipping of fresh salmon. Once transportation links were established, a booming business was underway.

> The salmon in the waters of British Columbia are excellent in quality, varied in species and most abundant. In the rivers, which they penetrate up to their head waters, they are caught by a drag net in the deep waters, and by a bag net in the rapids. In the sea they are generally caught with hook and line; a canoe at certain seasons can be filled in a day by the latter method. The Fraser River salmon is justly famous. They begin to enter the river in March, and different kinds continue to arrive until October, the successors mixing for a time with the last of their forerunners. There is a greater degree of certainty in the periodical arrivals of each kind in this river than at the coasts and islands. The salmon is used fresh, salted, pickled, smoked, and kippered, and for export is put up salted in barrels, and fresh in one or two-pound tins; the latter process has only been commenced during the past three years. The article produced is of a most excellent description, and will doubtless prove a source of considerable export trade when it becomes known in suitable markets. There would appear to be no limit to the catch of salmon, but the question of market

Columbia River salmon gillnetter hauling net and catch aboard (puller in bow and fisherman in stern), c. 1900. (HS)

must always be considered. (*Guide to the Province of British Columbia,* 1877–8)

...the fisheries of its seas...if properly developed, might be made extremely profitable; the fish, if caught and cured under European superintendence and with European means, might be exported profitably to Australia, where salmon and herring are both in demand, and the two extremities of the British empire might thus be made to join hands, with mutual benefit to each other. (Captain W.C. Grant, *Journal of the Royal Geographic Society,* 1859)

The salting process was first used on the Fraser by the Hudson's Bay Company at Fort Langley, primarily to provide a winter stock for employees and for local sale. As shipping facilities opened up, an export trade began and eventually reached large proportions. Chief Trader Archibald McDonald, who took over the fort in October 1828, asked his superiors to send him "a good Cooper," one who knew "something of fish curing." The shipment of salted fish became a regular feature of the fort's economy, and by 1838 it was thought the fort would supply all the salt provisions required for the coast. In fact it did more than that, opening the overseas commerce of the Pacific coast by shipping salt salmon to Hawaii.

While both springs and sockeye were traded, most of the output involved other species of salmon. Several private salteries were established along the river, centred around New Westminster; they shipped their product to the eastern United States, Australia, the Hawaiian Islands, China and Japan. The operation required minimum outfitting and was a business opportunity for men with little capital as well as those with more elaborate establishments. At least one of these salteries experimented with putting salmon up in cans.

The first cannery on the Fraser was built at Brownsville, opposite New Westminster, in 1871. It moved to New Westminster in 1873, and one or more other small canneries were in operation the same year. Statistics for the coast date from 1876, by which time there were three canneries. In 1883 the number had increased to twelve. It fell off to six in 1884 and 1885, but then there was a steady and rapid increase; there were 31 canneries in 1895 and 45 in 1898.

New Westminster was as far upriver as drift nets could be used. As the fishery moved downriver, the canneries followed, eventually concentrating along the lower part of the river, especially near Steveston and Ladner. By 1895 Steveston had half

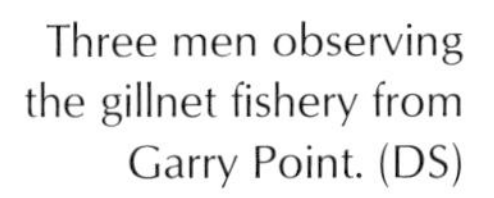

Three men observing the gillnet fishery from Garry Point. (DS)

Early daycare for workers at the Imperial Cannery, 1913. (F.D. Todd, BCARS)

the Fraser River canneries, and by the end of the century the line-up on Steveston's shoreline included Scottish Canadian, Atlas, Gulf of Georgia, Star, Federation (as ownership changed, it was also known as Steveston, Lighthouse and Empire canneries), London (also Lulu Island), Brunswick, Imperial, Hume's, Phoenix, Britannia, Pacific Coast, Colonial, Beaver (or Richmond), Canadian Pacific (or Red or Winch) and Garry Point canneries. "This place is now most centrally situated with regard to the more productive fisheries," reported the US commissioners, "having on one side those of the outer grounds and on the other those in the lower part of the river."

These early establishments depended on Victoria and San Francisco commission houses for their operating capital, canning equipment, canning supplies and export arrangements. Following designs used on the Sacramento and Columbia rivers, they were organized in assembly-line fashion with virtually every process (can-making, butchering, washing, cutting, can-filling, salting, cooking, sealing, weighing, labelling) done by hand. Most BC canned salmon was exported to England, generally in large lots. The rest was divided between Australia, other foreign markets and the Canadian trade. The crated salmon was taken by river steamer to Victoria where it was transferred to sailing ships.

The early years of the industry were prosperous ones. As demand for the product grew, more canneries were built, machinery was modified or invented to automate and increase productivity, new methods of capitalizing the industry were employed, and more fishermen in sturdier boats went farther afield in search of fish, using longer and deeper nets and sending their catches to the canners via collectors or tenderboats to prolong their fishing time. Larger quantities of fish were taken; there were often gluts that could not be processed in time and were thrown away as waste. As sockeye stocks diminished, canners started using other species to supplement their pack. By the turn of the century, freezing technology and a railway system to transport the product had been perfected; freezing needed salmon species with a low oil content to prevent rancidity. By this time Japanese immigrants had also started a fish-processing business, sending large orders of specially salted dry salmon to Asian markets. Today all species of salmon are valued by fish-

Chinese labourers soldering vent holes after exhausting cold air from cans. Scottish Canadian Cannery, c. 1900. (BCARS)

ing interests and exploited by the industry. It is, however, with the sockeye that the story of Steveston's glory years begins.

From the 1870s through the early 1900s, most canning operations were done by hand. A description by the British writer Rudyard Kipling of the Columbia River graphically illustrates the process:

> When our consignment arrived, the rough wooden boxes broke of themselves as they were dumped down under a jet of water, and the salmon burst out in a stream of quicksilver. A Chinaman jerked up a twenty-pounder, beheaded and de-tailed it with two swift strokes of a knife, flicked out its internal arrangements with a third, and cast it into a bloody-dyed tank...Other Chinamen pulled them from the vat and thrust them under a thing like a chaff-cutter, which, descending, hewed them into unseemly red gobbets fit for the can. More Chinamen...jammed the stuff into the cans, which slid down some marvellous machine forthwith, soldering their own tops as they passed. Each can was hastily tested for flaws, and then sunk, with a hundred companions, into a vat of boiling water, there to be half cooked for a few minutes. The cans bulged slightly after the operation, and were therefore slidden along by the trolleyful to men with needles and soldering irons, who vented them, and soldered the aperture. Except for the label...[it] was ready for the market.

In Steveston's first canneries, these operations were carried out by Chinese men and Native women working shifts of varying lengths, depending on the supply of fish. By the end of the early period, however, there was a flurry of mechanization. As more canners entered the industry, their capital stimulated invention and innovation. In addition, CPR construction was absorbing much of the province's manpower, causing chronic labour shortages in the fishing industry and a need for labour-saving devices. Finally, once an innovation increased the speed of one aspect of the process, the other canning processes had to be similarly accelerated to avoid bottlenecks in the canning line.

In the earliest canneries, each can was cut by hand out of sheet tinplate, formed and soldered. By 1890 a number of machines had been introduced to punch out body pieces, tops and bottoms, and to apply solder; but even when automatic can-making machines appeared at the turn of the century, many canners still preferred to make their own, believing it was no more expensive and knowing it gave a longer season's work to the Chinese crews they needed for processing the salmon.

In 1877 the first major innovation, the steam retort, was introduced. Basically a large pressure cooker, the retort could heat the product at temperatures higher than previously possible, thus speeding the cooking process. The retort decreased the incidence of burst cans, eliminated the problem of calcium chloride (a salt added to the boiling water in the old method) rusting the can, and required less labour.

Almost every piece of processing machinery was affected by the introduction of the retort. Soldering machines appeared in the late 1870s, and during the early 1880s innovations were introduced in the filling, salting and cutting processes, and in the movement of cans through the cannery. While these machines were not universally adopted until several years after their introduction—filling machines, for example, were not generally used on the Fraser before 1902 because the hand-filled product was neater and commanded a better market price—others, like the gang knife, a large knife operated with a hand lever with eight blades arranged to cut the fish into the exact length of a can, were important because they relieved bottlenecks in the canning line.

During this period most changes in the industry were confined to refinements in existing machinery. In 1891 a new method was introduced for packing salmon for the English market in half-pound tins rather than the traditional one-pound cans, which were subsequently discarded. There were two advantages to the half-pound tins. Just as much labour was required to put up the smaller cans as the pound cans, but in a season when fish were scarce the extra work was offset by the higher price obtained for the smaller cans. Also, the smaller tins did not need to be boiled or retorted as long as the pound cans, so more cases could be processed per day. One final innovation on the canning line before 1900 was the steam box. It eliminated the old system of double cooking in which cans were boiled in kettles for forty-five minutes before being vented and resealed for the second cooking in the retort. With the steam box, cans were passed back and forth through a box for only seven or eight minutes to exhaust the air.

As vital as these machines became, the industry would never have made it through the early years without the manual labour supplied by its Chinese workforce. While a few Chinese may have visited Canada's west coast in the heyday of the Pacific fur trade, the first true immigrants from China came to participate in British Columbia's gold rush in 1858, some arriving from California and some directly from Asia. In 1860 there were about 4000 in BC; the 1870 census counted 1500 Chinese. While the gold fields were the greatest attraction, some worked in Victoria selling provisions for the miners or acting as labour contractors. Others went into the growing and selling of fresh vegetables, cutting and selling cordwood, or operating restaurants and laundries. These immigrants were virtually all young men; most probably came with the idea of making their fortune and returning home. When the mining excitement died down they found work in other areas, such as domestic service, the coal mines and the Fraser River canneries. Hired on a contract system through "boss Chinamen," they provided a cheap and reliable source of labour.

The Chinese were welcomed by the canners. But the rest of white society felt threatened by such a large body of people who worked for wages that were too low to support a "white" lifestyle, who made little effort to learn English or otherwise assimilate with European society, and who seemed likely to take their earnings out of the country. White immigrants to BC often came from Australia and California, where there were strong anti-Asian sentiments. This attitude took root and grew in BC soil.

...because the Chinese, or the Japanese, through an evolutionary process which has

A Chinese cook outside a Steveston bunkhouse. (BCARS 77225)

been in progress for centuries is now, as we find him, a marvellous human machine, competent to perform the maximum of labour on the minimum of sustenance. He does not require to maintain a home as white men do; does not spend one 50th part of what the meanest white labourer considers absolutely necessary for clothing; lives in a hovel where a white man would sicken and die—and with it all performs...unskilled laborious tasks quite as efficiently as a white man, and, given the training, is equally proficient at duties requiring the exercise of some skill. (Victoria *Colonist*, June 18, 1905)

Initially Chinese people had full legal equality with other immigrant groups, but in 1872 the provincial legislature disenfranchised them. In 1878 the legislature, fearing a massive influx of Chinese workers to build the CPR, passed a law prohibiting them from being employed on any public works in the province. The federal government did allow the CPR's major contractor to bring in Chinese men, however, and between 1881 and 1884 about seventeen thousand came, some from China and some from the US. As the railway reached completion, some Chinese labourers returned home while others looked for work in Canada. At this time the Steveston canneries were being built at a great rate and needed large numbers of employees.

The white labouring class, who were also voters, feared a major upheaval in the province's work force that would throw them out of work. In 1884 a $10 head tax was levied on all Chinese immigrants. Once the railway was completed, the federal government raised the tax to $50 per new arrival, effective the beginning of 1886. The province banned various practices, such as the non-medicinal use of opium and the exhumation of bodies for reburial in China, which the Chinese regarded as an essential part of their culture. Chinese people were not eligible for Crown land. A federal royal commission investigated the closely packed conditions in Chinatowns and attempts were mounted to make the housing more expensive by establishing certain minimum cubic foot per resident requirements. The head tax was not entirely effective in halting Chinese immigration. The provincial economy, including the canneries, was growing. The $50 tax was not necessarily an impediment to a Chinese worker who could readily find employment. In 1900 the tax was increased to $100 and in 1904 to $500, and yet another royal commission, this time examining the issues surrounding Chinese and Japanese immigration, was appointed in 1902. Anti-Asian sentiment was successfully used in provincial election campaigns. The $500 tax proved an effective deterrent to immigration, but the Chinese who were already in the province soon realized their labour was in demand in several industries, including salmon canning. They used the labour shortage as a bargaining chip to demand higher wages. The canners reacted in two ways: they encouraged labour contractors to pay the head tax for new immigrants in order to increase the supply of workers, and they began using labour-saving machinery in the cannery, modernizing and speeding up the canning lines.

Many Chinese workers left British Columbia during World War One, but they returned in even greater numbers afterwards, expecting a postwar boom. The federal government reacted by replacing the head tax with the Chinese Immigration Act of 1923, a measure which effectively restricted Chinese immigration until after World War Two.

The Chinese people who came to work in Steveston were generally residents of the Vancouver, Victoria or New Westminster Chinatowns. They were gathered into work crews by a Chinese contractor, also known as a "boss Chinaman," working for one of the canneries. The contractor had an arrangement with the canner to provide sufficient labour to make the cans and:

> **take from the wharf and from scows alongside wharf, all tools and material landed there,...also take the fish from the fish boats at the wharf, prepare fish for canning, fill cans,...cook and properly test, lacquer cans so filled and label same, pack in cases, nail the same, and make ready for shipment, and that they will also do all repairing to coppers and other tools used and required by them in the execution of all the aforesaid work. (BC Packer's Association Chinese Contract)**

In return, the contractor was paid according to the size of the pack. From this money he paid the wages and room and board of his crew. In some cases the contractor may have paid the fare from China and the head tax for the labourers. He also set up work schedules and acted as interpreter and liaison between workers and canners. He might also be active in trying to find winter work for them.

Because it was inconvenient to commute between Vancouver and Steveston, Chinese cannery crews lived in bunkhouses provided by the canneries during the season. These "Chinahouses" were large, two-storied, barn-like buildings. They measured about twenty-four by forty feet. Half of the ground floor and all of the top floor usually consisted of rows of rooms, three bunks in each room. The other half of the ground floor was the eating area, the only heated part of the building. Cooking took place in a lean-to at the back of the building, in large cauldrons over brick fireplaces fuelled with cordwood. When canning operations were in full swing, the workers received three meals a day, usually rice, meat and vegetables; when things were not as busy they got only a morning and evening meal, all of which was arranged by the contractor. Reverend Appleyard's observations of one China House, recorded in 1902, reflect attitudes that were typical of the time:

> **Every cannery has its "China House", which is a large cheap-looking wooden building, designed to hold 40 or 50 Chinamen employed. It is often distinguished by the red dragon flag of China floating from a pole attached to the roof. The interior arrangements are a long narrow hall with cubicles opening into it from each side. The roof covers a bit of China, lopped off & transplanted on Western soil. Charms & prayers, made in China, decorate the smoke-begrimed walls; odours, the production too of China, occupy every square inch of space; low tables & benches are placed about the room, & the Chinese, squatting on the benches, reach the level of their food, which with chopsticks they sweep into their mouths; gambling & opium-smoking occupy the time not devoted to work, eating or sleeping.**
>
> **Come with us to the China House. No one knocks in China, so we go in without preliminary proceedings. The Indians, Japanese, & Chinese do not advertise their presence upon the threshold by beating the door...The expected "Come in" is the usual reply to a knock, but often the speaker's English is exhausted in saying it. Here we are in transplanted China: eyes, ears & nose bear evidence to the fact, the air bears a burden of odours anything but pleasant, it is a mixture of stale cabbage, rice, & opium, together with odours emanating from grease, steam fish, & perspiration with which the clothing of the Chinaman is saturated. To these various ingredients a leaky stove adds its contribution of smoke which fills the place. The rays of a lamp, hanging from a beam overhead, reveal the yellow-**

ness of the atmosphere, & the prayers, charms & fiends which scowl at us from the wall. It reveals also the forms of men grouped around a table, reminding one of a herd of driven cattle as they sway & push against each other in their eagerness to see the gambling game which is being played. The illusion is aided by the long tail which hangs down each man's back & swings with the motion of their bodies. (Rev. Appleyard)

When the season ended most Chinese workers returned to the larger cities or contracted themselves out in other industries or occupations. Sometimes a small group was able to make ditching contracts with farmers. The farmers provided tents and a stove; the labourers brought their own food, cooking utensils and bedding. Some Chinese eventually bought farms in the Steveston area. A few cannery workers remained in the Chinahouses over the winter, existing mainly on rice, the cabbages they had grown over the summer, and dried fish. Scurvy was often a problem. If a Chinese worker died during the season, a coffin was made from lumber found around the cannery. Personal items were packed in and the empty spaces filled with rice before the lid was nailed on and the coffin placed on the wharf to be shipped elsewhere for burial.

While in Steveston the Chinese tended to live, work and socialize with their own group; there was not much interaction with the whites, Natives or Japanese except during the course of the work-day. Leisure time was spent in the Steveston Chinatown, originally at the foot of Second Avenue and later at the foot of No. 1 Road. Both sites were on the waterfront and fire eventually destroyed both. The foot of Trites Road, where a number of Chinese stores and services were located, later became the popular gathering spot. Probably the best known was the Hong Wo general store, built in

Hong Wo and Company, a store in lower Chinatown. (CRA Bell-Irving)

1895 by Ling Lam, who became a Steveston resident. The Hong Wo (meaning "good luck" or "living in harmony") store was built on piles over the Fraser's banks near Nelson Brothers Cannery and the old Imperial Cannery. Fishermen could tie up outside. Ling Lam devised a system of distributing grocery/hardware checklists to the fishing fleet so they could make up their orders in their spare moments and have them filled while they waited on board. At first he commuted to New Westminster for his supplies; later he used the telephone to call in his orders. Ling Lam's business prospered and he was able to buy one hundred acres of farm land next to his store, where he grew tomatoes in greenhouses and potatoes, beans and cucumbers in the open fields. He built eight- foot concrete vats for pickling the cucumbers. After 1914 Ling Lam became a Chinese contractor and built bunkhouses on his property for cannery and farm workers. The area around his holdings became known as a gathering place for Steveston's Chinese community and evolved a spirit of its own. When he died in 1939, his funeral parade through Steveston was well attended.

Drinking, gambling, going "with the hookies" and opium smoking were how the Chinese occupied their nonworking hours. Many scrupulously saved every cent, however, looking forward to their return to China. Some tended gardens next to their bunkhouses. Opium smoking was a custom brought over from China and was legal until 1908, when the manufacture and sale of opium was banned in Canada. Before that, Chinese merchants in Vancouver, New Westminster and Victoria openly cooked and sold this product.

> **Buster McKenzie: Used to have water pipes. Big octagon pipe about that size, and there was the little connection coming out of it where they put their stuff in and then you suck the smoke through the water. They had the water down in a bucket and they'd draw on it. What they put in it I don't know. All the camps had those water pipes...**
>
> **Barbara Heeren: And then we used to see them [the Chinese cannery workers] on the weekends. Sometimes they would get a day off, but that was after the cannery was working. And they would be lying just in heaps of humanity. My brother and I used to go there and peak in, with their long opium pipes [at] Hong Wo's, the next place up. They were slaves really, the Chinese were, not the Japanese ever, but the Chinese were, and they worked in the cannery. With these long pipes, we didn't know. We just know that they were asleep was how we saw it. Except one of them would lift his head and go [shows puffing a pipe], you know, and draw this on his pipe and that's how they spent their day off, if they ever had one.**

Opium was once readily available in Steveston, in tiny bottles like this one.

> **But we always went to look. [They sat] just on the floor, mud floor, just lying there, all up against each other. Well, of course we know now what it was. But I guess that it was opium, we were told that they smoked after we were adults.**
>
> **Jimmy Hing: In those early days there were quite a few Chinese around here. There was pretty near 1000 of them. There was about three or four opium dens around here, the Chinese people smoke opium you see, around Steveston here, in the area. Say within two square miles they have the opium dens. Sometimes they have one in the corner there, or underneath that building [the rancheria] there, when it was over back there [by the dyke]. It was on stilts about that high. The people, in the summer time, they put those Chinese, what do you call, bamboo, rattan sheets, they lay on there and smoke opium you see. So the police won't see them. (*Steambox, Boardwalks, Belts and Ways: Stories from Britannia*, 1992)**

The Chinese were not ignored by the missionaries, who took advantage of the summertime concentration of workers to spread the Gospel:

> **Circumstances have this year prevented a continued stay of any of our missionaries to the Chinese at the canneries. But we were able to arrange for five three day visits. In the last week of the season I was at liberty to accompany our missionary, Chan U. Tan to the grounds. I had the pleasure of half an hour's conversation with Dr. and Mrs. Large. Part of the afternoon I spent in getting acquainted with the leading Chinese shop-keepers. Water street, in Steveston, for the most part, consists of the narrow summit of a dyke. The walking along it, and the feat of passing people without tumbling off takes one back to China and Chinese highways. Some six thousand Indians, Japanese and Chinese were about the place. Our evening meetings for the Chinese are held in the open air in the Chinese quarters. I noticed that we were not able to gather anything like the crowds we do at such meetings in Victoria, Westminster and Vancouver. The people are very excited over their fishing and gains, and it is not possible, as at other times, to collect them; nor to hold for any time those who are gathered. All about is a laughing, frolicksome crowd...In China public readers of novels and plays are an institution. Our workers adapt the idea in Steveston. During the day the missionary takes his place in one of the eating houses and reads aloud from the New Testament to the people. No doubt much good seed is sown in this way. (D.S. Moore, *Methodist Recorder*)**

The Chinese were employed exclusively as cannery workers; they were not fishermen. Before the American Can Company automated the business, their work year started in February or March, making cans for the coming season's pack. They used shears to cut the tinplate into can bodies, then stamped out more forms for the tops and bottoms. At the same time, the solder cook was preparing solder bars from pig lead and pig tin for use in the machines that fastened the seams of the cans. When enough pieces had been punched out, a "former" began rolling can bodies by feeding the tin sheets into a set of rollers. The seamers placed the rolled tin on a cylinder, brushed the seam with muriatic acid, placed a small bit of solder on the seam, and went over it a time or two with a soldering iron kept heated on a charcoal stove. When the melted solder had secured the seam, the form was passed on to men who fit the can bottoms, then used a crimping machine to press the flared edges of the bottoms against the can bodies. The edges of the bottoms were painted with acid, then passed through a solder machine to complete the process. The cans were stored until they were needed. It was something of a gamble to decide how many cans to make. Surplus cans could not be stored for use the next year for fear of rusting. On the other hand, an insufficient supply meant the canning lines had to be stopped mid-season while a new batch was formed, and the down time was costly.

Once the fishing season began, the Chinese worked at several types of jobs on the canning line, with varying wage scales. Some were slitters who cut off the head, fins and tail, slit the belly and cleaned out the entrails. They worked at long tables arranged in a U-shape with wooden screens to prevent splashing their co-workers. Fish were dropped through a small opening in the screen into a sliming tank with running salted water. Slitters scraped the waste product into a trough that dumped into the river under the canneries. This practice was later prohibited and the waste went into hoppers under the canneries; when the tide was right, the waste was emptied into scows that were towed by a cannery tug to the deep water of the Gulf for dumping. Federal fishery regulations later banned this practice too, and the scows began transporting the waste to an oil reduction plant on a nearby island.

The next group of workers to handle the fish were the slimers, who scraped off loose scales and more thoroughly cleaned out the inside of the fish. They then dipped them in salted water and put them in a hopper to dry. Both these jobs (slitting and sliming) were shared by Chinese men and Native women. The next step was always done by the Chinese. Fish were cut into sections, by hand-operated gang knives in the early days and later by power-driven revolving knives. The sections went into another hopper to drain off excess moisture and then were brought to the chopping blocks where expert butchers quickly and accurately split each section into the appropriate size for the cans. The butchered pieces went into wooden tubs.

Butchers were the most highly skilled and well paid of the cannery workers. The speed of the butchering process determined the speed of the entire canning line in the early days of the industry. Butchering crews were generally composed of about thirty men, each of whom processed about 1,500 to

2,000 fish in a ten- hour day. Labour shortages caused by the head taxes and the resulting demands for better wages by those who remained in the industry were great incentives for the canners to consider mechanizing the process. In 1905, a peak salmon year, canners were faced with widespread labour shortages. In 1906, the Smith Butchering Machine was introduced to counteract this problem. The first model, nicknamed the "Iron Chink" because it eventually replaced the Chinese butchering crews, processed sixty to seventy-five fish per minute with the aid of three men. The first machine butchered the fish only after the heads and tails had been removed, but by 1907 it had been modified so that it cleaned the entire fish automatically. Apart from being faster and cheaper, it increased the profit per fish by decreasing waste and giving a consistent quality of butchering.

Filling was usually done by Chinese workers. Standing on either side of a long table with a groove running down its middle to hold the butchered pieces, fillers put no more than three pieces of fish into each can, with the dark skin next to the tin. A little salt was added to the can before it came to the filling table. The filled cans went on to the wiping tables, where more Chinese workers would use old, cleaned netting to clear off any debris that might have attached itself to the cans. Cappers put a small piece of tin on top of the fish and then put a top on the can. The tops were made with a vent hole and the projecting edges of the vent hole rested on the piece of tin so that the fish wouldn't clog the hole and block the escaping steam as the cans were soldered. The tops were tapped into place with a piece of wood, then crimped to the can bodies, then soldered. After soldering, the vent holes were also stopped with solder.

Cans then travelled to the bathroom for cooking, again with Chinese crews. This was another key point in the process, where the capacity of the cooking equipment determined how much fish was processed. The contents of all the cans had to be cooked that same day. Bathroom crews were often the same men who had made up the can-making crews before the fishing season. They first put the cans into hot-water testing kettles. Any cans that produced bubbles, indicating a leak, were removed and resoldered. The cans then went into cooking kettles for an hour, immersed in boiling water heated by steam pipes strung in the bottoms of the kettles. Cooking kettles eventually were replaced by steam boxes, which could be heated to temperatures above that of boiling water, reducing the cooking time by half. After this first cooking the bathroom crew punched a small hole in the top of each can with a nail and mallet to let the steam escape. The hole was stopped with solder and the cans were again tested in hot water. As before, any cans leaking bubbles were removed, resoldered and retested. The final

1906-1907-1908-**1909**-1910-1911-1912

THE BIG YEAR
THE BIG OPPORTUNITY

ARE YOU PREPARED?

CHINESE LABOR WILL BE SCARCE
CONTRACT PRICES WILL BE RAISED

Every cannery in the Sockeye district will operate. The far-sighted packers are preparing now. Are you one of them? Don't you know Chinese labor will be scarce and that the Chinese contractor has already advanced his price?

Protect yourself by installing an "Iron Chink" and make him give you a labor-saving allowance. He is now allowing enough per case to pay from 10 to 20 per cent per annum on the cost of the machine, depending on the locality and the number of cases to be packed. That is big interest on the investment, not to mention the saving of the cost of the "Iron Chink" in the *fish saving* and again in the *increased capacity* as we have told you in our previous advertisements in the Pacific Fisherman.

The run may last only 12 or 15 days and by using an "Iron Chink" you can pack your full retort capacity. Get your cans full in the shortest possible time, and close down and save expense. Don't wait until the season is over and then figure that you could have packed as many cases as your neighbor, if you, too, had had an "Iron Chink,"--it will then be too late.

You have only been practicing during the last three years, but NOW is your opportunity to make a clean-up.

Are you going to neglect it?

Forget the cost price of the "Iron Chink" and figure results. We have doubled our output capacity but from present indications we will have to work overtime to supply the demand.

If you wish an early delivery, you must order now.

Smith Cannery Machines Company

SEATTLE, U. S. A.

This advertisement for the Smith Butchering Machine, known as the Iron Chink shows the racial attitudes of the time. (*Pacific Fisherman*)

cooking was done in retorts, essentially large pressure cookers. After cooking, the cans went for yet another hot-water testing, then they were washed with a hot caustic soda solution, hosed with cold water and taken to a packing room. When they had cooled, a Chinese tester made a final inspection. Tapping each can with a tool that had a ball-shaped head and a handle like a spoon, the experienced tester could tell by the sound if the can had a leak or was underweight. Any can failing this test was labelled a "do over" and was not included in the pack. English importers had complained bitterly about the high percentage of spoiled salmon in the original shipments and specified in their contracts the penalties they would impose if "do-overs" were shipped.

At the end of the season cans were labelled with the canner's own labels, again by Chinese workers. Sometimes they were sold to distributors unlabelled.

> **From the retort the crates of canned salmon were transferred to the stock room. I noticed that one thousand cases were not labelled. On enquiring I was told that by arrangement with other canneries they would remain unlabelled until it appeared as to what brand was the market favourite. (Alfred Carmichael, *The Emigrant Boy*)**

Label paste was made of flour, water and glue, brought to boil and cooked until it reached the right consistency. The labeller sat on an empty salmon box with another box in front of him and one by one attached the labels. As they worked their way through the mountains of cans, other workers packed the cans into wooden cases and nailed tops on them. The cases went to a warehouse area near the cannery's wharf to await shipment.

At the beginning of the season each Chinese worker received a tag with a number, which he was to wear in a visible spot—some tucked it into a hat band or pigtail. The bookkeeper was stationed at the door of the cannery and each time a worker entered and left the building (lunch, quitting time) the bookkeeper recorded his number and the time of day. The bookkeeper then had to figure out how many hours each employee had worked that day. Timesheets were posted on the Chinahouse dining room hall by noon the next day so that any irregularities could be dealt with immediately, usually with the assistance of the Chinese contractor. After disputes were settled, times were entered into a China Labour Ledger, with a separate page for each worker. Payment was made at the end of the season.

As machines like the Iron Chink and the retort increased the speed of butchering and cooking, other processes were mechanized as well. Gradually, can-making, filling, salting, washing, wiping, weighing, capping and all the other steps on the canning line were automated; by 1913 the modern canning line was established. Since then, only a few refinements have been necessary.

The Chinese workers complained bitterly that the introduction of these innovations put more and more of them out of work. But the canners were not about to back down. Their new methods not only were faster, they also produced a more consistent quality of product with fewer "do-overs".

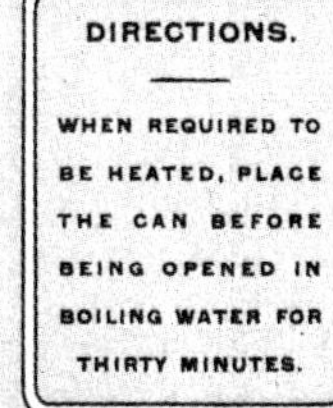

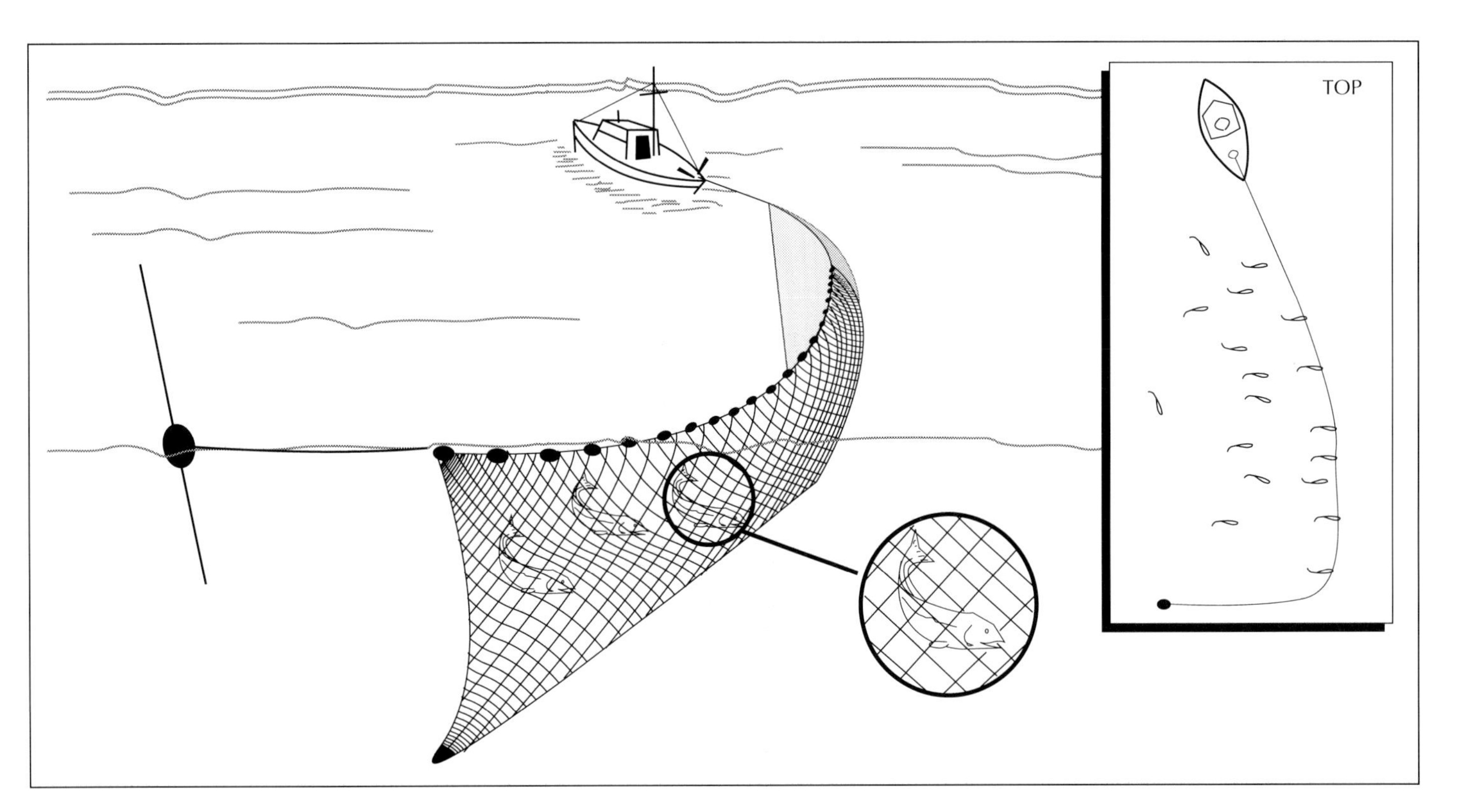

How gillnet fishing works. (Lionel Trudel)

Canneries could not do entirely without their Chinese crews, however. They continued to employ them on the contract system until the Chinese cannery workers became unionized in the mid-1940s.

In the early years the Fraser River canners relied solely on the gillnet, or ensnarement, fishery for their supply of salmon. It appears that gillnets were not used by the Natives, who favoured entrapment techniques such as reef nets, dip nets, traps, weirs and spearing. Entrapment fisheries, however, were too slow and unreliable to ensure a large enough catch for continuous cannery operations. The centre of the Native fishery was in the lower Fraser canyon, too far from New Westminster, the hub of the immigrant industry, to guarantee the catch was suitable when it arrived at the cannery.

The Hudson's Bay Company was experimenting with different types of nets at Fort Langley as early as 1829, so it is entirely possible that some long-forgotten company employee initiated the use of the gillnet on the Fraser River. Many Fort Langley employees were from Scottish islands where gillnet technology had been used for centuries.

The gillnet's dominance was due largely to the Fraser's opaque waters during the sockeye season, which enabled the gillnet to be used effectively during daylight or at night. In addition, early conservation regulations discouraged or banned encirclement (seining) and entrapment (traps). Finally, the Fraser area has protected waters in which relatively small, low-cost vessels (compared to trap and seine vessels) can operate. The inexpensiveness of the original gillnet vessel, or skiff, was especially important in the early years when capital was meagre. The decision to limit the fishery to gillnetters also created more jobs than would have been needed for a trap fishery.

A gillnet is a piece of webbing with a mesh that allows a fish to pass through as far as the gills but no farther; the fish is caught by the gills. It differs from a seine net, which catches the fish by surrounding it. The size of the mesh varies according to the species of fish sought. The netting is loosely suspended between a lead line on the bottom and a cork line on the top. These nets are fished either by anchoring them as set nets (illegal in BC, but not on the Pacific coast of the US) or by allowing them to drift with the tide or current as drift nets. Since it takes fish in its individual meshes, the net may be payed out in a straight line, an arc or any other desired shape. Its dimensions are usually given as length in fathoms and depth in number of meshes. The effectiveness of this type of gear depends on enough silt being in the river to obscure the net, since the fish will avoid an obstacle if they can see it. In the early spring the Fraser is comparatively clear, so fishing is carried on only at night. Sediment appears in the river later in the season and is at its greatest intensity during the months of June, July and August, after which the river begins to clear again. In opaque waters the net is used effectively day or night. It is during this season that the great sockeye run, on which the early canneries depended, takes place. This gear was used not only in the river itself but beyond its mouth where the dis-

Gillnet loft, c. 1925. (BCP)

coloured water extends for several miles in all directions. The Fraser's gillnet grounds stretch from Point Roberts to Point Grey to a distance of at least five miles offshore and up the river above New Westminster by a few miles. Beyond the Royal City there were few good drifting places during sockeye season and less certainty of a sailing breeze, an important factor when the gillnet vessel relied solely on oar and sail power.

The original Fraser River gillnet, used until the 1890s, was double knotted and made of soft twine. The twine was made of loosely laid flax and had a very coarse appearance. Flax was used instead of cotton because it is stronger and deteriorates less quickly when used in a river fishery. Unlike cotton, the holding strength of flax is greater wet than dry.

Before knitting machines capable of producing the double knot were introduced in the late 1880s all gillnets were hand knitted, mostly by Native women.

> **In addition to these they employ a number of Indian women knitting nets; in length these nets will average from 120 to 150 fathoms, at a cost of one dollar per fathom. A better idea can be obtained of the amount of work required to make one of these nets, when we reflect that their length is about 900 feet, with an average depth of 16 1/2 feet; about fifty of these are kept continuously on hand, sufficient to meet any mishap that may occur, as they are constantly wearing out and otherwise being injured; the amount of money required to sustain this department in full working order can be estimated from the fact that one net will stand but one month's constant fishing; it is then thrown aside as useless. ("Our Salmon and Salmon Canneries," *The Resources of British Columbia*, 1883)**

The total cost of a gillnet (mesh, lead line and corkline) was between $120 and $150. Bluestone (copper sulfate), cutch (tree bark from the Philippines) and tar were used as net preservatives. The cork or float line was of hemp and floats were of cedar, though in some cases tin cans were used. The Fraser's lumber mills produced lathed cedar floats until a net float factory was established in 1889. Early gillnet leads were simply attached to the bottom line, not woven into the lay of the rope like modern lead lines.

Between 1871 and the early 1900s there were very few changes in the dimensions of the gillnet except for net depths. In the early 1880s conservation measures limited the mesh size and the maximum length of sockeye nets, but until a 1908 regulation limited net depth to sixty meshes, gillnet depths were set by custom, which was in turn governed by the depths of the major drifts. Before conservation measures took effect, the nets had become increasingly deep as the fishery gradually moved out of the river and into the Gulf, growing from twenty-seven meshes in 1883 to one hundred or more meshes in 1900.

At the start of a fishing day, the cannery's net boss saw that the day shift fishing crews got out to the grounds.

> A varied nationality is represented among the drift-net fishermen, including Indians and negroes, there being a very large number of the former. The arrangements with them differ. Some own their boats and nets and dispose of their catch by contract; others are supplied with their outfit by the canneries and fish on shares, while others again, the Indians especially, are employed on day wages. The independent fishermen in possession of an outfit is supposed to fish it himself, and his hours are measured by his endurance. The canneries, however, generally hire two gangs for each of their boats, in order that they may be kept at work both day and night. The licenses do not define the position which each fisherman may occupy with his drift net. The law provides, however, that the nets shall be kept at least 250 yards apart and shall not be used so as to obstruct more than one-third the width of the river, but it has been manifestly impossible to comply with these regulations—the first, especially—since the number of nets has increased so greatly; and the second, because in many places the width of the river is less than three times the length of the nets.
>
> The fishermen are left to arrange these matters among themselves, and whether they do so by tacit understanding or not, there is little or no interference among them. Each selects, so far as he can, what seems to him the best location, and may change it from time to time. As the nets are floating no fisherman has a clear piece of ground to himself, but they follow one another in groups over the same ground, and move upstream again after completing their drift or after having made a certain distance. (US Commission on Fish and Fisheries, *Report of the Commissioner for the Year Ending June 30, 1899*)

A cannery tug towed the boats out to the fishing grounds, and at the end of the day the net boss wrote out the details of who fished and how much they delivered. The tally man counted the catch as the salmon were thrown onto the wharf. Fishermen often tried to break the boredom of the job by targeting various spots or people on the wharf. In the early days, when each salmon was worth only a few cents, accuracy was not critical. As the stocks declined and the fish increased in value, the job became more demanding.

Gillnetting was done in a "reach" or a "drift," a stretch of water fairly uniform in depth, and free of snags or sharp ledges that could catch the lower portion of the net. Before gillnetters were mechanized, each boat had a two-person crew. The puller rowed the vessel and the fisherman took care of the net. In setting the net, usually an hour before high water slack when the salmon head upstream, the puller rowed across the current while the fisher-

Salmon gillnetters of the Columbia River style, setting off the mouth of the Fraser River. (NAC 88-20945)

man payed out the net. When two-thirds of the net was out, the boat was turned downstream at nearly right angles to its former course so the net ended up in an "L" shape. The boat drifted with the net until an hour after the turn of the tide or until the end of the drift was reached. The net was then hauled over a wooden roller on the stern and the catch was removed. This procedure was repeated if the tide was not too strong and if there were not too many boats working the drift. A contemporary account describes the process as follows:

The net used is 150 fathoms long, which is put in the water and allowed to drift with the tide for about half an hour, when it is taken up and thrown out again until the time for returning is up, or the boat is full. The boats go out at six o'clock in the morning and return at six in the evening, when another shift of men take their places and fish from six at night to six in the morning. They frequently drift four and five miles from shore, and have caught as high as 800 fish to the boat. (Vancouver *Daily World*, August 28, 1889)

The boats in the first commercial fisheries on the Fraser were dugout canoes and flat-bottomed skiffs. Flat-bottomed skiffs appeared as early as 1870 and continued to be in general use as long as fishing stayed in the river itself. They were powered by sail and oar: "The sail helps to get you from one area to another...but within an area you were manoeuvring with oars" (Jack Downing, in *Draft for the Marine Development of Richmond*). However, "when the wind failed it was 'out oars and pull' and rowing a heavy boat or dory is back breaking work

Sail-driven Columbia River gillnetter delivering salmon to a pot scow in the 1890s. (CVA)

at any time" (*Canadian Fisherman*, February 1914). The cost of a skiff during this period dropped, probably as a result of mass production. Initially they cost $46, but by the 1880s they were as low as $31. These skiffs, commonly known as Fraser River skiffs, were generally twenty-foot, flat-bottomed double enders, with big flares to the side and a round bottom fore and aft. There were slight variations in skiffs on the upper and lower reaches of the river fishery; boats on the lower river were more heavily built and more extensively rigged as protection against open water conditions in the river mouth.

Fraser River gillnet fleet, c. 1900. (DS)

Fishermen originally worked drifts located very near or at the cannery. As they went farther afield, however, another type of vessel—the steam-powered tenderboat—was introduced to transport fish from the fishing grounds to the canneries. The earliest evidence of their use is 1877, when the Leonora and the Leviathan collected fish from English and Company boats on the Fraser and took them to the processing plant. There were two types of tenders: freighters and tugs. The freighter carried fish in boxes on the deck or in the hold, and the tugs towed the fish in scows. By 1878 their importance to the industry was indisputable.

Soon after the introduction of tenders, changes were made in the method of collecting salmon. Rather than having tenders obtain them directly from fishing boats, fishermen unloaded their catch into scows at fish camps. Twice daily the tenders picked up the scows and towed them to the canneries. The fish camps, first used around 1881, consisted of fishermen's living accommodations and net racks for repairing nets. At first they were both shore stations and float camps, but the floating variety proved superior because the shore station serviced only one section of the river while the float camp could be towed to the best fishing grounds. Camps were established at favourable fishing points along the river at the beginning of a season, with Native fishermen and netmen under the charge of a white man. Boats needed four men apiece, since fishing was done in two twelve-hour shifts.

The fish camp and scow system improved on the original freighter system of collecting fish, especially in turnabout time. A freighter had to wait to load and unload, whereas a tug simply exchanged an empty scow for a loaded one and serviced a greater fishing area in less time than a freighter. Tenders also increased the mobility of the gillnet fleet by towing fishing boats to and from fishing grounds. This was especially useful when one area of the river had poor fishing and boats needed to move quickly to a better location.

Some fishermen owned their own boats and gear, and fished under contract to a cannery. Not wishing to be dependent on contract fishermen, however, canners also had their own boats and gear for which they hired fishermen, initially Natives, later Japanese. Before the season started boats were repaired, then painted with the cannery's colours. A tug captain could readily identify which of his company's boats might need a tow to another drift, or which fishermen might be selling fish independently instead of delivering them all to the cannery. The paint also helped to quickly reveal stolen or lost boats.

The Native workforce often came from distant areas for the fishing season. Natives preferred working for a day wage because it enabled them to move on to other seasonal work, such as hop picking. This proved a disadvantage for the canners who needed a steadier work force for the short, intensive, often unpredictable salmon season. For this reason canners much preferred to hire Japanese fishermen when they became available, because they were prepared to work on a contract system, guaranteeing an entire season's work. They were also more productive. Canners continued to hire Native fishermen, however, because they needed their families for cannery work in the era before the automated canning line had been developed.

The aboriginal residents of the Fraser delta belonged to the Coast Salish Nations. Before European patterns of life were established, they inhabited more or less permanent winter villages and migrated to other areas to harvest foodstuffs, where they lived in temporary shelters. The Coast Salish had spring camps for the herring and shellfish harvest, then moved on to summer salmon fishing sites, the rights to which were owned by families. When the canneries were established in the 1870s, the Natives were the largest population group in the province and a natural source of labour for the fishery.

> **The Indians were fishing in this region when it was first invaded by the whites. They were then, however, solely concerned in supplying their own domestic wants, using apparently the same appliances they do to-day, reef nets and hooks and lines in the salt water, and spears, dip nets, and weirs in the rivers. Traders reached the upper Fraser very early in the century, thence working to the sea, and the salmon became one of their most important foods, being obtained partly by their own efforts and partly of the Indians. The latter gradually developed into commercial fishermen, and to-day constitute a prominent element in the fishing fraternity. (US Commission of Fish and Fisheries)**

The Hudson's Bay Company employees at Fort Langley used to buy salmon from the Fraser River Natives for their fish salting operations, but this fishery was too far upriver to reliably supply the first canneries located in the New Westminster area; if fish did not arrive at the cannery within a few hours of capture it was unsuitable for canning. As canneries moved downriver, there was some thought they might be supplied by the Native fish traps near Point Roberts, but this did not occur to any great extent. When the salmon runs filled the Native traps, they also filled the nets of boats working closer to the canneries, and by the time the Natives delivered their catch there was already a surplus of fish ready for canning.

As the canneries established themselves, Natives from all along the coast—including Vancouver Island, the Queen Charlotte Islands, the upper reaches of the Skeena and Nass Rivers, and Washington state—moved their families and goods to the cannery sites and contracted to provide labour.

> **The packing of salmon for food is one of the leading industries of British Columbia, affording employment to thousands of people and yielding an annual income of several million dollars. The Fraser is the greatest salmon-producing river, and on its banks there are many canneries. On this mission there are sixteen of these canneries, and Steveston is the centre of operations. In the winter season it is a small village of 250 or 300 souls, but in the summer it has a population of over 5,000, representing nearly every nation—about two-thirds are Indians, mostly pagans. (Rev. A.E. Green, *The Missionary Outlook*, January 1895)**

Some canneries had Native labour recruiters. Chinook jargon, the old trade language, was used to overcome the difficulty of working with so many language groups. Natives who had a good summer at one cannery tended to return the next year. However, if they could afford to be selective, they travelled from cannery to cannery, checking out offers and working conditions, before committing themselves to an employer.

The Washington state Natives generally arrived on horseback; others came by canoe.

> **But spring is nevertheless heralded by the coming of Indian canoes. In April & May every tide bears its contribution of Indians...With the Indian comes, not merely his family, but every live thing he possesses; dogs, cats, fowls, & more! & receive frank hospitality in his own dwelling place; these visitors, fish, flesh & fowl, bring life & movement & add an element of activity which effectually wakes the slumbering villages. (Rev. Appleyard)**

Depending on the weather and their starting point, it could take ten days to a month to reach the Fraser. Once the Natives had contracted with a cannery, their canoes were stored, sometimes at the end of their row of houses, sometimes on the river bank. Large canoes were left outside the dyke; the smaller ones were brought into the camp, turned over onto logs and covered with rush matting.

The Native families lived on a portion of the cannery property specially reserved for them,

Natives arriving at New Westminster, 1887. (Notman Archives, McCord Museum)

areas called rancheries with cabins or camps. The houses were built in a row of one-room accommodations, usually made of split cedar or sawn lumber with roofs of cedar, matting, scrap tin or tin plate from the cannery. The back of the dwellings often served as primitive outhouses, and were the subject of many public health concerns. Houses were occupied only during the summer fishing season (though they were a source of fascination for the children of Steveston the rest of the year), and if the previous year's occupants returned to the same cannery, they had first claim on their old house. The buildings were furnished with whatever the Natives had brought with them for their summertime needs.

> The floor of the interior of these one-roomed houses is littered with blankets, furniture, cooking tins, fish gear, carnival masks, and usually three or four dogs...
>
> The Indians brought some of their own native foods: sundried salmon, probably a holdover from last year, this may sustain life I refuse to classify it as food for humans; their smoked salmon and oolichans were quite good; the really valuable food for them was oolichan oil which had all the medicinal value of cod liver oil. They used it as freely as we would use butter or even more freely. I am of the opinion that the great incidence of TB which took such a heavy toll of life was due to their abandoning this valuable food for our butter and animal fats. They brought it along in quantities, mostly in coal oil cans (5 gal). They used to buy what we whites used as bedroom crockery—jugs, basins, soap dishes &c but they were for kitchen and table use by them—why not? Anyhow to see on their table a jerry pot half full of weak tea colored oolichan oil did anything but simulate the white mans appetite. (Garnett Weston)

Sometimes the Natives erected smoke houses for smoking fish for winter food; these were posts holding up a roof, with rush matting part way down one side as a windbreak.

Right across from the livery stable, behind it, running where the alleyway is just between, say, that little hall and the Steveston Hotel, well, from Second Avenue to Third Avenue there was a long building and all just single rooms with a door, a verandah over, and a post or so—each family had one room, you see, that was for the Indian families. It was double, one lot facing south and the other facing north. It was demolished and they had a big two-storey building right down here on Garry Point for a number of years. (Harold Steves, Sr.)

The houses were marked by numbers that corresponded to the numbers on the fisherman's cannery boat, the wife's spot on the canning line, and the family's account at the cannery store. Each person had a square of tin perforated at one end and impressed with his or her number. They threaded string through the perforation and tied knots in the string to keep track of how many days they worked.

In the early days of the industry, before the extensive involvement of the Japanese, the Native workforce was essential for the operation of the canneries. Their arrival was an event of great importance along cannery channel.

By the time I was sixteen I was thoroughly conversant in Chinook, and it was this knowledge that got me my first job with the cannery at Steveston in 1899. The manager was reluctant to give a young student a job until he heard I could speak Chinook, as it was important to have someone in the cannery who could talk to the Indians who came to work at the cannery, all of whom spoke Chinook. I was also employed at Steveston to interpret in the police court when the regular interpreter was away (drunk)...

O.K. All is ready for the salmon run but the run is still a week or 10 days away and the Indians are not here yet. But we are ready. . . But the Indians wern't [sic] here yet and they were essential to the operation as the women

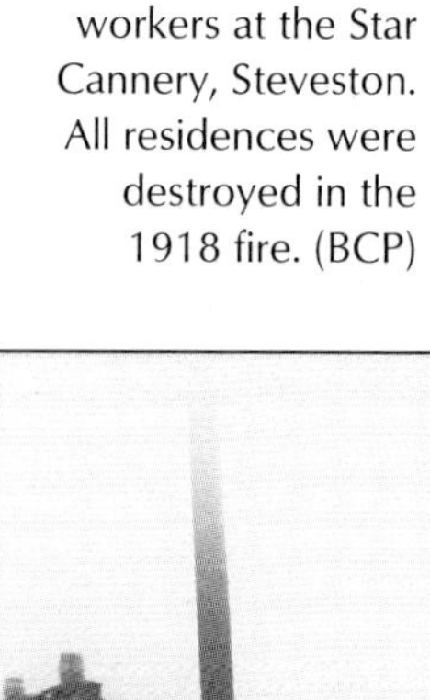

Housing for Native workers at the Star Cannery, Steveston. All residences were destroyed in the 1918 fire. (BCP)

Native living quarters, 1913.
(F.D. Todd, BCARS)

did all the work at the sliming tanks and practically all the filling.

Well here they come on a flood tide in the late afternoon bowling along ahead of a westerly wind. Great canoes 50 ft and over, spread to an 8 ft beam, each with 4 sails, wing & wing. Some came all the way from the Skeena and the Queen Charlottes but most were Kwakiutl ie Cape Mudge to Queen Charlotte Sound. They came in a flotilla about 12 of the big ones and an equal number of lesser ones about 40 ft or a little less. I'm sure one of the big canoes could hold up to 100 if you counted children but did not count dogs, mongrels all, short on pedigree but long on multiplicity of breeds. Most of them showed various stages of distemper or mange; they were quarrelsome and forever fighting.

There were plenty of children as we found out in the next few hours. About 1/2 the flotilla kept on up the river to other canneries.

There was considerable ceremony to the landing. A few hundred yards out sails were lowered and they started singing Indian songs until they got to 5 or 6 yards from the wharf. With a final shout a sudden silence. Dead silence for about a minute. Then the biggest chief stood and with a speaking staff in his hand made a short speech. A sudden stop. A barked order. And then all hell broke loose: every paddle clawing water to sidle the canoe up to the wharf; mooring lines made fast. The lower 5 or 6 feet was under water but we had planks ready to bridge it to the first canoe & then more planks over the first and second canoes to make a causeway for the outer canoes.

To us it looked like utter confusion. Everyone seemed to be rushing here & there with something whe [sic] dropped somewhere & ran back for more. Kids slithering through from the canoes in waves one after another. The first to come up were in pairs—a Siwash & Kloochman (I use these words as they were used at that time. They were soon to be deemed uncomplimentary and dropped into complete oblivion) each carried a load of iktas (things). These were dropped in a handy spot & the man took off for a truck or dolly. Then back to the canoe for more of the family all bringing more iktas. Up and down till their complete outfit was on the wharf. I didn't see anything more from the wharf. When the kids came up and started rambling around completely out of control I high tailed it back to guard the lye vat.

Let me go back a bit. When word was flashed around that the canoes were in sight it was an order for all hands on deck to keep the cannery from being wrecked. Everything movable was put away or at least well out of sight. Alex came and as he passed the lye vat he saw that it was full. "Is the lye in that yet?" "Yes." "Well who in hell did that? Why is there no cover on it? Never mind then, DeBeck! Guard that vat. Don't let anyone come near it. If the kids or anyone ever got their hands in it they would be ruined."

Thats why I left the wharf on the double. You can hardly blame the kids for getting wild with excitement. Remember—they had been living in a canoe for a week or more, wedged in amongst their iktas, no water for washing, bathroom facilities over the gunwale. The last reach of the trip from Seymour Narrows to the Fraser about 130 miles they always tried to make in one run. And they generally did it. Prevailing winds at that time of the year were westerly, generally starting in the early forenoon and dying at sundown. Sailing they could make over 5 knots with a brisk wind but paddling only about 3 plus. So they would be sailing about 12 hrs and paddling 12. In other words nearly 2 full days to make the run (I learned all this by talking to the Indians). With all that pent up energy it was natural that the kids would get out of control.

Native children at a Steveston cannery. (BCARS)

The lye vat was a heavy sheet metal vat or pan 6'x 6' about 18" high set on the floor and nearly filled with a nearly saturate solution of lye, which is pretty lethal water.

I didn't have long to wait. A group of 5 or 6 little ones came up on the run. I jumped in front of them "Klatewa (scram) Hyak (quick) Thats death water in there. It will kill you if you touch it. Klatewa. Oh Hell here's more of them—Bill! Charlie! Lend a hand here. Get a club & help keep the little bastards out of that vat" All this shouting of mine was in good Chinook and good English. The kids understood but they didn't heed worth a cent. Their dirty little faces registered disbelief and their brains were figuring that if this needed a special guard it should be further investigated; it might be good to drink. "Get someone to get the firehose out" and in Chinook "Hey you there Peter Moses Daniel or Paul or whatever your name is get these kids away from here. If they get their hands in that water it will burn their hands off"

Word was passed back and up came a couple of Klooches. When I explained to them it was soon over. A good clout on the side of any head within range started them all back to help with the unloading. (I use the word "help" with reservations.)

I got no more customers. So far as guarding the vats was concerned I could just as well have been sitting in an armchair reading poetry.

My station was at the back of the cannery a good 100 yds from the wharf where the confusion was. I could see where 4 or 5 chinamen were still stamping out covers. I saw the raid on the waste tin was nice shiny sheets each with 12 big round holes in it. True—each sheet had cutting edges and needle points. There was a mob urge to collect resulting in quarreling and grabbing; then a line of them going back dripping blood for first aid. This was of two kinds either a piece of dirty rag tied around it with string or a spanking and instructions to get down there and bring up more things from the canoe—each equally effective.

Fitz passing through looked over at the charcoal stoves. "Now what in hell do those little devils want with charcoal out of those stoves? No! dont stop them. Tell them to help themselves, fill their pockets take some for their brothers & sisters. Show them the boxes where they can get all they want. Tell them its good to eat. It's cheap, we've got lots more."

I didn't see much more, but I heard much when one removes half a dozen tins from the lowest stratum of a 6 ft high pyramid Newtons law comes into play and there is a whole avalanche of tins—noisy. Also you can't climb a column of stacked trays without having them come down over you. We had all these & more. Confusion, yes, but there was order behind it and finally everything was unloaded and stacked in mounds on the wharf. Alex gave the order to take what was needed for the night; doors would be locked in half an hour and they could get the rest tomorrow.

The men moved the canoes on the high tide as close to shore as possible. Tomorrow they would be pulled up into the brush and carefully covered with cedar bark mats to keep them from cracking in the sun.

Next day they were back at sunrise to move their iktas to their houses. Then men who were to fish in cannery boats had to be assigned and given their little book for the tallyman to enter each delivery of fish and initial it with corresponding entries in his own book. The women to get their places at the sliming tanks or filling tables—each one numbered. (Edwin Keary DeBeck)

Generally the men worked as fishermen and the women worked in canneries washing fish and filling cans; they also made and mended nets. Sometimes the wives and daughters of the fishermen worked as boat pullers, rowing and manoeuvring the boat while the fishermen tended the net. Children who were old enough to be useful sometimes worked on the canning lines, cleaning and lacquering the filled cans; they might also find work in the can lofts. Young children generally stayed with their mothers on the canning line; the women worked with infants strapped to their backs.

The cut-up salmon was then delivered in buckets to benches at which the Indian women stood or sat and neatly fitted the pieces into cans...If the Klootchmen had babies, which was quite often, they sat at their benches with the babies on their laps. They would suckle them or attend to their toilets, then continue filling the cans with salmon. I protested to the manager that the process was hardly sanitary. He told me not to mind—"It will be all sterilized in the cooking," he said. (Alfred Carmichael, *Account of a Season's Work*)

The Natives generally used boats, nets and other gear belonging to the cannery and received a daily wage of $2–$3. They fished under licences held by the cannery. The boat puller was usually paid about 25 cents less than his or her partner, the fisherman. The fishing and canning season on the Fraser only lasted from the end of June until early September. At the end of the season some Natives returned to their villages; others went on to harvest work, particularly hop picking in the Fraser Valley and Washington state. Before leaving Steveston they often stocked up at the local stores, buying a wide variety of articles, some of which were later distributed at potlatches.

While the fishing and cannery work was a departure from their traditional life cycle, many ele-

1828—INDIANS, EWAN & CO'S CANNERY, NEW WESTMINSTER

Native workers, Ewen's Cannery, 1887. (Notman Archives, McCord Museum)

ments of aboriginal life endured. The leadership of the community still revolved around people from the villages; Chief Capilano is noted for having visited the cannery where his people worked. Many residents of Steveston remembered the singing and dancing in the Indian camps near 7th Avenue. Gambling was a popular pastime for the men, particularly poker, three-card monte and their own gambling games. Only European sensibilities were offended by the fact that the women earned extra money through prostitution:

> **The Indians...lived in cannery cabins on shore. Half of them were women with many children. Their way of life was very different. I wouldn't say they were immoral but rather, amoral. It was not considered wrong for a woman to make a dollar at the oldest of all professions. (Edwin DeBeck)**

Missionaries followed the exodus from the villages to the summer workplaces and made a special effort to be in Steveston during the fishing season.

> **In no other portion of the province is there such a gathering of those who must need to be reached by the Gospel, and living for several months so near together that it is easy to reach them in large numbers, while to try to reach these same Indians when scattered in their own villages would necessitate hundreds of miles of travel, months of time and great expense. (Rev. A.E. Green, *Missionary Outlook*, January 1895)**

Missionaries played another role. Having a foot in both cultures, they realized that Natives were being treated less favourably than white fishermen. They helped them become aware of the value of their services to the canners and to negotiate better contractual arrangements. Missionaries tended to the practical needs of the Native people, acting as intermediaries with various authorities:

> **Our Indians have shown their appreciation of the missionaries services by splendid church gatherings from the fifteen tribes scattered along a coast-line of over one thousand**

Natives eating a lunch of canned salmon at the Imperial Cannery, 1913. (F.D. Todd, BCARS E5064)

miles...Also, from the worldly standpoint, our Indian friend has shown his regard in entrusting his interests with us. Their requests are legion. By way of entertainment we relate the diary of August 21st, confining ourselves to calls only.

1. A young Indian asks us to assist him in securing the consideration of the Inspector of Fisheries for a special licence to retain his Indian weir or fish-trap, the heirloom of generations. 2. Another comes with a sorrowful tale of money lost. 3. A young girl complains of the Chinese contractor cheating her of her honest wage. 4. Matrimonial harmony disturbed. However, the wife does not seek alimony but simply her share of the earnings she and her husband gained in the fishing. 5. Applications for written statements to present to the several customs officials on their entering Uncle Sam's domain to pick his hops with their dainty, dusky fingers. Ten applied to-day. "Still there's more to follow." 6. Visited a camp and tried to assuage the feelings of friends grieving over having run amuck of the Children's Aid Society, the latter preferring to assume the charge of a little unfortunate half-caste. 7. Acted as general accountant in reckoning value of salmon caught by Indians, and hours of labor done by their spouses. 8. Attendance at court, having acted as interpreter to make the Indian prisoner's mind clear as to the magistrate's duties in his regard. 9. Visiting the sick, distress having been brought on by impure sanitation. 10. Another of hymen's misfortunes, ending in mutual regrets and possible separation. (Rev. William Stone, *Missionary Bulletin*, 1905)

Two Native women enjoy ice cream as they stroll along the boardwalk, 1913. (F.D. Todd, BCARS 84162)

By the turn of the century most canners had adopted the system of paying for fish by the piece rather than by a seasonal or daily contract. The guarantee of a certain amount of money was replaced by the uncertainties associated with an unreliable supply of fish and variable fishing conditions. In addition, Japanese fishermen, who entered the fishery during the last two decades of the nineteenth century, were slowly but surely crowding the Natives out of the fishery.

Native women working in the canneries were often hired and paid by the Chinese contractor. Some women were regular employees who repaired nets; others were strictly seasonal staff, the spouses of the fishermen employed by the cannery. The average working day was ten hours, but it could be longer at the height of the season. In the early years of the industry, before gillnet webbing was made by machine, Native women were employed before the fishing season as knitters. They were given a certain quantity of linen twine to be made into the mesh size required by fishery regulations and to the depth desired by the cannery. They held a wooden gauge (to ensure standard size) in their left hand and their shuttle or needle filled with twine in the right. Each net took a few days to complete, after which it was measured; if satisfactory, a new allotment of twine was issued.

During the canning season the women worked mainly as slimers and occasionally as slitters, fillers or can wipers. While their mothers worked, the Native children might pile the filled salmon cans or play on the beach under the cannery floor when the tide was out. Here they often found knives and whetstones that had fallen through cracks in the floor, and they sought rewards for returning them.

While they were at the cannery houses, Natives cooked over open fires, either inside the building or out. Many varieties of fish made up their principal food source, including salmon, sturgeon, oolichans, and shellfish such as clams and cockles. When there was a surplus of salmon at the canneries, canners offered the extra to the Natives, who cured it in their smoke houses or salted it in a strong brine. They also preserved oolichans. There were special treats for children, such as the uncooked spinal cord from the sturgeon their parents occasionally caught in the Fraser. They also made a simple bread from flour and water, baked in an open pan, and their meals were accompanied by butter, greens, wild rhubarb and berries gathered from fields around Steveston.

Clothing was generally simple and convenient. Women wore dresses that could be pulled over the head without need for fasteners. A friend had the woman stoop over, laid a newspaper on her back, and cut the outline of her figure, which was then transferred to fabric purchased at a Steveston store. Young children often wore nothing more than a cotton shirt, and neither they nor their mothers were accustomed to wearing shoes, stockings, coats or hats.

Men and children sometimes swam in the river, but women rarely did and thus were more prone to vermin infestation. Missionaries visiting the Native camps helped procure medical help for other conditions as they became evident. Deaths were common in summer camps and the Natives frequently wanted to take the bodies back to their reserves at the end of the season. They made coffins from lumber found around the canneries and stored them, sometimes in trees or small "houses," until they went home. Vandalism brought an end to this practice, and they started taking their dead to Vancouver; but the custom provided a vivid memory for many Steveston pioneers.

The Natives always brought their dogs along with them and sometimes, at the canneries, made pets of hair seals who lived at the mouth of the river and were the bane of the fisherman's life, stealing salmon from the nets or becoming entangled in them. European fishermen generally knocked them over the head and threw them back into the water; Native fishermen often brought them back for the children to play with.

> **The seal was fed salmon heads from the cannery waste, and within a few days it became tame enough for the children to play with, cautiously at first, but with increasing familiarity. It eventually became tame enough to be taken out in a small canoe to swim playfully about with the children. It frolicked in and out of the canoe with them until the canoe was swamped, and then the children and the seal tumbled out and returned to the camp, the seal squirming its way up the bank.**
>
> **Two sounds I have never forgotten are the seal's cry at night and the song of the Indian inducing fish to swim into his net. The cry of the seal closely resembled that of a dreaming infant; the Indian's song was a composition of his own—a tune of a few bars, sometimes with improvised words. On a quiet moonlit night, with the stillness so absolute as to be heard, the cry of the seal or the song of the Indian broke through with a suddenness both weird and startling. (T. Ellis Ladner, *Above the Sandheads,* 1979)**

In the early years the canners were dependent on the Natives for both fishing and cannery labour. As the Japanese entered the industry, however, canners preferred them as fishermen, finding them more productive and less likely to take sudden leave of the canneries in favour of other occupations, such as harvest work. When Steveston suffered the double calamity of losing several canneries to fire and losing large salmon runs to a variety of environmental changes, the Natives stopped coming in great numbers during the salmon season, staying closer to their upcoast homes where other canneries were being built. By 1918 only one Steveston cannery used Native labour in any significant number.

Steveston's summer population was a unique ethnic mixture, which was a distinctive feature of the little farming village and a source of great fascination to outsiders.

"Strange Fascination"

The date of my visit [1901] synchronized with the great annual rush of the fish from the sea up the river to deposit their spawn in the flowing water that has had the chill taken off it by the hot summer sunshine. Helter-skelter, crowding together with an eagerness to obey natural law which is as cosmic as the daily rising and shining of the great luminary on the soft Pacific waves, they come with a rush...

We drove seventeen miles to Steveston to see the canning. There are twenty-nine canneries at this queer town of plank streets, wooden houses and big canneries that straggle all along the river front. It is alive and kicking for two or three months in the year—the rest of the time it sleeps, and the visitor then wanders through deserted thoroughfares and shut-up canneries. Our drive took us...to a region of fertile farms, off which a big hay harvest had just been reaped; and at last, over a mile and a half of the straightest and dustiest road it has ever been my lot to travel, to the long straggling Steveston canneries...

The canneries are great sheds opening on one side of the river; and going into their semi-darkness out of the sunshine we became conscious of the performance of a great provision industry by the agency of blue-bloused, sallow Chinamen of all ages. First, however, let us walk to the river front, where the boats are coming in with their cargoes of fish. Soon we are looking out on the bright-glancing, drab water—for the Fraser is turbid. There are many sailboats—sixteen or eighteen feet long, perhaps; broadish in the beam, strongly built. They go out into the river, and from them nets are cast, into the meshes of which the eager fish rush and are caught by the gills. The principal work this year seems to be taking the fish out of the nets and knocking them on the head to stop tail-wagging.

Twelve and a half cents each the men get for their fish. A moment or two ago at the hotel you had seen one of them give the landlord $250 to take care of for him. He said he had earned it during the previous week.

We turn to the interior of the cannery, and find ourselves standing by a great heap of salmon. The heap is about forty feet long and ten wide. There may be ten or fifteen thousand salmon there. Flanking this is a long, wet, fishmongery-looking bench, at which a solemn old Chinaman is at work with a sharp knife. It is rather wet and slippery everywhere, so that you walk carefully and there is such am anphibious atmosphere about that you might expect to meet a mermaid.

But watch Allee Knifee. Is it by magic that the fish's head comes off with a clean cut; then his tail; then his big back fin, and his side fins? The head is pitched through a hole in the board that fronts Allee Knifee, the tail through another, the body through a third. Facing the old master of the knife, on the other side of the board, is a long tank of water, and here a dozen Indian women stand cleaning the fish—for Indian women clean salmon by immemorial instinct. Chinese labourers are perpetually carrying the cleaned fish to a most ingenious machine, that goes by power, and saws the salmon up into lengths just the height of the ordinary salmon tin.

It is entertaining to watch the thin circular saws, gauged to the proper distance apart, on the axle, cut up the fish into short lengths, and roll it down a shoot. A score of Chinamen and Japanese are packing these short lengths of salmon into cans; rolling the red fish into cylindrical form, and jamming it into the receptacles with much art. The cans are then weighed, and if they do not contain the prescribed quantity, more fish is jabbed in. These full cans are then carried in trays to another most ingenious machine into one part of which they are fed, while into another part a supply of can tops is kept going. The machine puts the tops on the cans; that is, it pushes them on tight. If one of them misses getting its cover, a Chinaman or a Japanese seizes a top and pushes it on by hand.

Still, however, they require soldering. For this another clever arrangement is put into requisition, whereby a succession of cans roll down an incline. They roll a-tilt, on the circular edge of their tops, the cylindrical portion of the can being at an angle of forty-five degrees. They also roll through a thin stream of molten solder, which is kept hot by fires underneath it. Before they start to roll they pass under dropping acid (technically known as 'fake') which acts as a flux for the solder. They emerge with their tops soldered on air-tight. Then they are placed in a boiling vat for an hour and ten minutes. On coming from this ordeal a small hole is pierced in the top of each can, which allows of the escape of the imprisoned steam and moisture. This hole is immediately soldered up again and packing is complete. When the cans are cold they will be japanned, labelled, and packed in cases.

There is a strange fascination about this haunt of industry—this assemblage of continuously working Celestials and Japanese and stolid, broad-faced Indian women. There is no chatter; you scarcely hear a word spoken from the time you go in till you come out, for strict attention to business is characteristic of these Mongolian workers and their confreres. But here and there, leaning up against a bench, is a well-worn tin opium pipe...

It was lamentable to see the great overplus of salmon at the canneries this year. I saw heaps of fish that were refused by the canners simply because they had a greater supply than they could pack...At present large quantities of fish have to be thrown away or used as manure. The abundant catch of this year means much to Vancouver. It is so much wealth cast up by the sea into the outstretched hands of these British Columbia workers. (Bernard McEvoy, *From the Great Lakes to the Wide West,* 1902)

Workers preparing and salting salmon, 1940s. (VMM)

Workers on the herring canning line, 1940s. (VMM)

Cannery workers placing herring in a gibbing machine, 1940s. (CFC)

Workers hand filling salmon tins, 1940s. The canneries often used twins in their public relations photographs. (CFC)

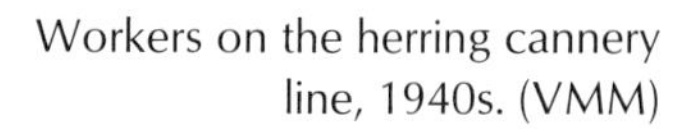

Workers on the herring cannery line, 1940s. (VMM)

Chapter Four:
The Fishery Becomes Big Business

Fraser River gillnet fleet moored at Star and Gulf of Georgia canneries. (VPL 26788)

In its early years the fishery was small and the resource was enormous, and there was little concern about exhausting it.

> There is no law governing fisheries in British Columbia. Fishing is carried on throughout the year without any restrictions. This state of things is well suited to a new and thinly populated country. The restrictions of a close season would be injurious to the Province at present, and for many years to come. (*Guide to the Province of British Columbia for 1877–8*)

Not many years passed, however, before regulations were imposed as conservation measures.

The first federal Fisheries Act was passed in 1868. No special provisions were deemed necessary when British Columbia joined Confederation in 1871, since there was no evident threat to the fishery. When it became obvious that the salmon canning industry was a thriving concern, the act was extended to BC in 1876. Regulations specific to BC waters were first enacted in 1878, confining salmon drift nets to tidal waters; forbidding salmon nets of any kind in fresh waters; forbidding salmon drift nets to obstruct more than one-third of the width of the river; and prohibiting salmon fishing from 8:00 a.m. on Saturdays to midnight on Sundays. Subsequent amendments to the Fisheries Act for BC were designed to promote conservation, regulating mesh size and net depths, and prohibiting seines, traps and so on. Another measure was a weekly close time for the fishery:

> **All fishing being prohibited from 6 o'clock Saturday morning until the same hour Sunday evening, the salmon are given an unobstructed passageway up the river during thirty-six hours out of every seven days. The movement of the fish is not, of course, uniform or even continuous throughout the season or any extended part of it. While, therefore, it is impossible, without the necessary observations, to pass definitely upon the matter, yet at the end of each weekly close time it is expected that a proportionally much greater quantity of fish may be found in the neighbourhood of New Westminster than at other periods of the week. On Sunday evening, as the time for fishing reopens, the work begins actively about New Westminster, the river being covered by as many boats as can safely operate, and the catch per net being as good as at least the average on the lower drifting grounds. Such success does not continue long, and during the remainder of the week comparatively few boats remain on the upper grounds. In the interest of the protection of the fish it would be important to ascertain what proportion of the run is removed by the large amount of netting used on the Fraser River during the past few years...the evidence presented by the circumstances attending the weekly close time argues strongly in favour of the continuance of that protective measure. In illustration of this matter may be cited the catch by the drift-netters during the night of Sunday, August 16, 1895, which was said to have exceeded 700,000 sockeye, the largest single night's catch on record up to that time at least. (Richard Rathbun, *US Commission of Fish and Fisheries Commission Report,* 1899)**

Legislation did not guarantee compliance with its provisions, and on the Fraser, where enforcement for most of the year was left to one guardian with one small launch, the law was ignored. The guardian had the help of two extra agents in the peak fishing months of July and August, but they were so poorly paid only local men who needed seasonal work could afford to take the job. Too vigilant an enforcement of the laws brought on the wrath of the canners and other neighbours on whom the men depended for other work. Cases were heard before local officials who tended to sympathize with the lawbreakers. The report of a detective hired in 1894 to investigate fishery conditions on the Fraser documented widespread abuse ("the present act being a dead letter, owing to its never being enforced"). The investigator made several recommendations including: appointment of a superintendent with jurisdiction over all matters relating to Marine and Fisheries; increasing staffing and patrol equipment; permanent appointments for fishery guardians; abolition of patronage appointments by federal MPs and creation of a more professional class of guardians; repeal of the general fisheries act and its replacement by specific acts for each river, lake, stream and estuary; the licensing of fish processing plants as a means of forcing their compliance with the law; and the fixing of minimum and maximum fines for breaches of the regulations to give the local magistrates only a small leeway in their dispensing of justice. Detective Galbraith's work received considerable attention in Ottawa, but the federal government took no direct action. They did, however, use several royal commissions to keep informed about the BC fisheries in order to formulate future regulations and policy.

By the end of the nineteenth century the fishing industry was taking a heavy toll on the salmon resource. Other industrial activity was also putting pressure on salmon stocks. Logging, mining and roadbuilding often destroyed spawning grounds. Canals and ditches for agriculture diverted spawning fish from their upriver migration, and smolt from their journey to the ocean. Agricultural

Salmon fishing on the Fraser River, view from the Scottish Canadian Cannery. At centre is a sail-driven Columbia River boat; at right is a motorized Columbia River boat.

refuse fouled salmon streams. Sawmills discharged fish-killing substances into rivers, and even canners fouled the Fraser with fish offal. No one questioned the fact that the salmon supply was rapidly diminishing. Fraser River canners, distressed and united by the decreased numbers of fish, urged the provincial government to take control of the fishery in a manner similar to the system in the US where individual states controlled the fishery resource. Encouraged by an 1898 Privy Council decision that allowed the province to levy taxes on its fisheries and use the money for the betterment of the industry, canners offered to contribute, through special taxes, such sums as were necessary to make the fisheries department self-sustaining, with the understanding that revenues would be spent on the protection, preservation and betterment of the salmon fisheries. In 1901 the province responded by creating a Fisheries Department which exercised authority over the fishery for the next twelve years. In 1913, however, the Privy Council confirmed a 1912 Supreme Court decision that the right to regulate fishing in the province rested solely with the Dominion, a decision that curtailed much of the work undertaken by the province. Subsequently the department confined its work to a study of conditions affecting the fisheries; inspections of the salmon spawning areas; determination of the characteristics of the salmon runs to each area; studying the life history of the salmon, and issuing annual reports for the benefit of the salmon industry. Funding for this work came from fishery taxes and licences.

Provincial and federal horns locked again in 1919 when the Minister of Naval Services issued a new schedule of fishery licences for BC, advocated limitations on the numbers of canneries and gear, and announced that revenue from these licences would be used for various fishery improvements. The province questioned the right of the Dominion government to place a tax on fish canneries, while the federal government defended the move as a regulatory measure. The Supreme Court eventually ruled that the privilege of granting licences to canneries was vested in the province, but that this did not give the province the right to regulate fishing; it simply declared that fish, when caught, became personal property within the province and thereafter subject to taxation.

Meanwhile the resource, which had gone from nonprotection to regulation by two levels of government, continued to suffer from the stresses of industrial development, especially in the 1899–1903 and 1910–1917 periods. During the first period, damage was associated with a dam at the outlet of Quesnel Lake. Built without an adequate fishway, the dam nearly annihilated the run into the lake. During the second period, railway construction through the Fraser canyon dumped rock into the river, blocking the progress of the salmon. The Hell's Gate slide of 1913 has often been cited as the single catastrophic event that nearly destroyed the Fraser fishery, but the destruction was more likely the cumulative effect of railway construction.

Salmon runs were now so diminished there were fears the commercial sockeye industry was finished. Fraser River salmon were also subject to fishing pressure when they passed through American waters, and an international treaty was seen as the best method of regulating fishing to restore the runs to their former abundance. The Sockeye Salmon Fisheries Convention (which later had its powers extended to include other salmon species) was signed in 1930, though ratifications were not exchanged until 1937. Its provisions cover all the territorial waters and high seas westward of Canada and the United States between Bonilla Point, on Vancouver Island, and Tatoosh Island, in Washington state, including the Fraser and its tributaries. Regulations enacted by the International Pacific Salmon Fisheries Commission are enforced within their territories by each national government. By the terms of the treaty, the commission was charged with making a thorough investigation into the natural history of the Fraser River salmon, into hatchery methods, spawning ground conditions and other related matters, and with conducting the sockeye fish culture operations in the waters under its jurisdiction. It was also empowered to limit or prohibit the fishery (though in no way infringing on the right of either government to regulate the issuance of fishing licences) and to prescribe the size and meshes of all fishing gear used during the various fishing seasons, to guarantee a proper escapement of fish to spawning grounds. The work of the commission has been successful in rebuilding salmon stocks in many areas and in protecting them against the pressures of the fishing and other industries.

In general, however, there was no master plan for fisheries management in the early years of the canning industry. The commercial "users" and industrial "abusers" came into conflict over who had rights to use and regulate the fishery, and these problems were faced only as they arose—and often deferred for years while jurisdictional disputes were settled—leading to a piecemeal approach to policy making on provincial, federal and international levels. An example is the response to fishing pressure on the Fraser River in the late 1880s.

Since the productivity of a cannery rose with the number of boats fishing, it is not surprising the fishing grounds on the Fraser were overcrowded by the late 1880s. Between 1872 and 1888 the number of canneries increased from three to twelve, with increased output on each canning line as well. As early as 1881, most fishing occurred at the river's mouth and the sandheads as competition drove fishermen to seek grounds closest to where the fish first appeared. In 1884 over four hundred boats fished on the river and by 1888 serious overcrowding was reported:

> **If all the canneries are in operation I do not understand where the room for the increase of nets is to come from. At the regulation distance apart the number of nets fished this year would extend 85 miles, while there is only about 70 miles of fishing ground. (Canada, *Sessional Papers,* 1889)**

The federal government introduced fishing licence limitations between 1889 and 1892. During 1889, 1890 and 1891 the number of licences was limited to five hundred with an average of twenty licences per cannery. In early 1892, however, in response to strenuous protest from canners and fishermen alike, all limitations on the number of boats were lifted. All bona fide fishermen who were British subjects received licences, and lax handling of applications made evasion of the British citizenship clause very easy. Canneries and other establishments dealing in salmon were allowed several nets each; independent fishermen were entitled to one net each. Their boats were marked with their licence number. Canneries and dealers had separate series of numbers as each received a single licence for all its boats.

The lifting of the licence limitations led to overcrowding of the inside grounds and resulted in many fishermen moving out into the open waters of the Gulf of Georgia. Two years after the removal of licence limitations, the number of fishing units increased by nearly 50 percent. By 1891 the fishery extended to the edge of the clear Gulf waters, as far as it could go with existing boat and net designs.

A new type of gillnet, less visible in clear water and thus better suited to the Gulf, was introduced in 1892. Constructed of oiled, hard laid sturgeon twine, the new net not only increased the range of the fishery, it also extended fishing time because it did not bunch up and become entangled as easily as the soft twine net.

The original Fraser River gillnet skiff was also found to be unsuitable for the increasing intensity of the Gulf fishery in the early 1890s; it tended to split in two in heavy open waters. A new vessel, the Columbia River boat, was introduced between 1889 and 1892, although it may have had limited use earlier. The *Canadian Fisherman* described the gillnet boat of the early 1900s:

> **The boat is a strongly built round bottom sailing boat, 30 feet long, 6 1/2 foot beam with a 6 foot centre board. On either side of the centre board are the fish tanks capable of holding two and a half tons of fish line weight. The sails are a jib and an ordinary sloop rigged main sail. (July 1917)**

The sails were wrapped around the mast, and the mast lay along the gunwale.

> **The design of the boat was controlled by the need to carry a great load, yet it had to be of such size that it could be rowed. The strong winds met on the Pacific Coast dictated a small, hand rig. The simple rig...was reduced in a squall by unshipping the sprit and tying the peak to the mast hoops to form a small triangular sail. The boom could also be unshipped easily...The boats were built with**

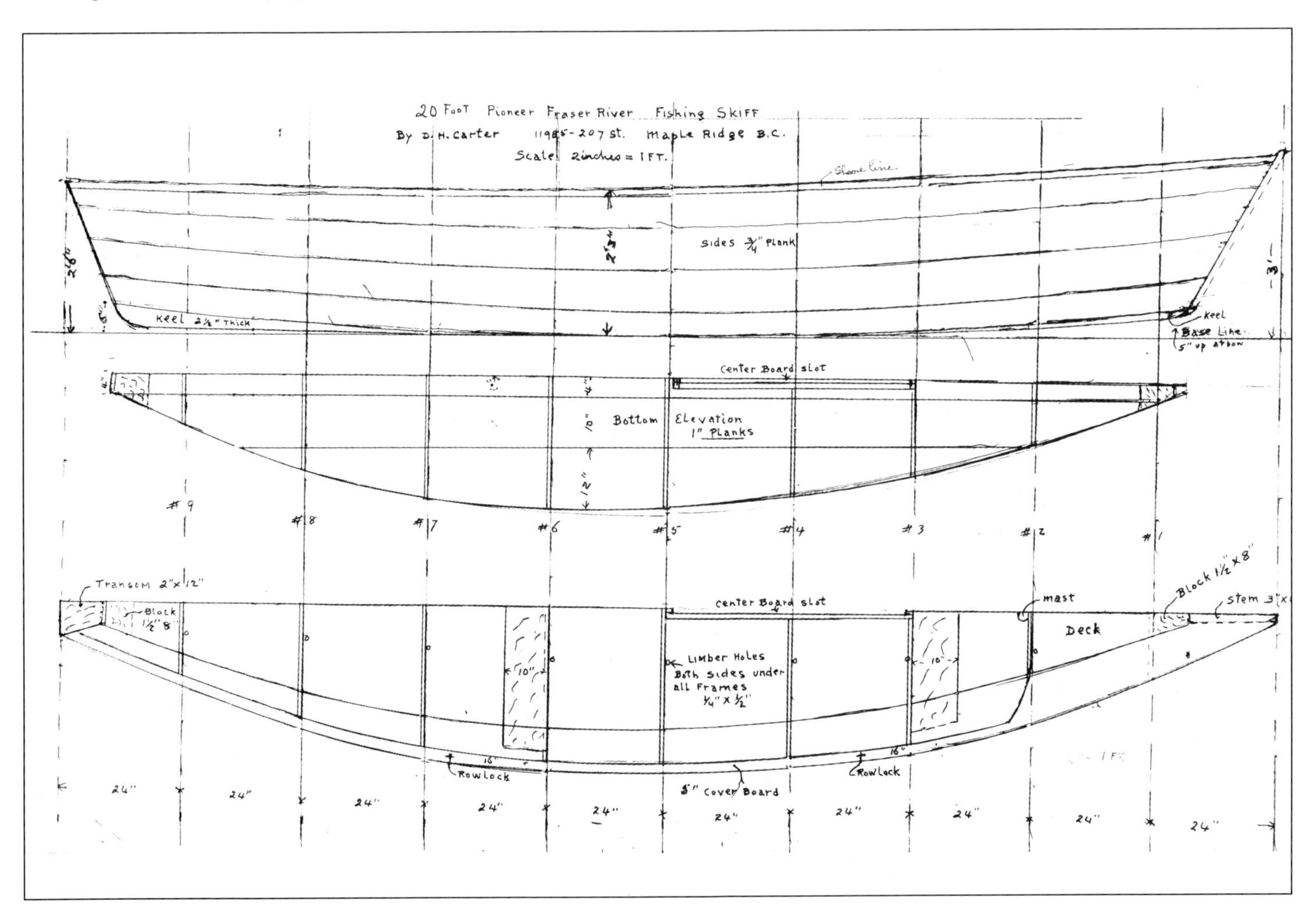

Atagi Boatworks, 1911. (Atagi family)

steam-bent frames, which were first bent over molds and later placed in the boat, cold. Planking was Oregon fir (Douglas fir, Oregon pine), and larch was sometimes used for the keel and the posts. The boats were built in a sound manner without frills. In spite of the hard usage inherent in their employment...they are said to have had an average working life of from ten to fourteen years. The construction was quite conventional in all respects and was moderately heavy. The Columbia River salmon boats were not intended to be very fast sailers, yet in fresh winds and with a load they showed a good turn of speed. (Howard Chapelle, *American Small Sailing Craft,* 1951)

Although such boats were commonly referred to as Columbia River boats, the name is misleading. Another boat built for the Fraser River fishery, the Collingwood boat, had a similar, if not identical design. The "Columbia River boat" may have been the West Coast name for the Collingwood boat, which had been built on the Great Lakes by William Watts, possibly patterned after Irish boats. The design was transplanted to the Fraser by Watts's son, Capt. William Watts, who established a boat yard in Vancouver in 1888 to construct gillnet boats for the Fraser River canneries. The first Columbia River boat used on the Columbia was said to be built in 1869 in San Francisco. Since William Watts Sr. built boats as early as the 1850s, his design may have travelled to San Francisco where it became known as the Columbia River boat—or it may have had a parallel development in the US. The Columbia River boat was used earlier on the Columbia than on the Fraser before 1899, when migratory fishermen came from the Sacramento and Columbia rivers to fish the Fraser each year, so the name probably came from these migrants. In 1891 another shipwright, Andy Wallace, who had worked for William Watts in eastern Canada, moved to Vancouver and shared gillnet boatbuilding contracts with Captain Watts before establishing his own yard in 1894 and becoming the area's principal builder of Columbia River boats.

Initially, Columbia River boats were constructed by European shipwrights. Between 1895 and 1897, the Wallace Shipyards alone built from six hundred to eight hundred of them. Between 1896 and 1901, however, about half the manufacture of fishing boats passed into the hands of local Japanese boatbuilders. There were two reasons for the rapid inroads made by the Japanese. First, non-Japanese boatbuilders who did not mechanize their yards could not compete with the low-cost labour provided by the Japanese. In Wallace's shipyard, wage rates ranged from $1.25 to $4 for a nine-hour day, whereas Japanese builders would work ten and eleven cents an hour, or ninety to ninety-nine cents a day. (Wallace successfully competed with the Japanese by applying labour-saving devices and woodworking machinery.) Second, the Japanese contracted with canneries to build the boats and also guaranteed to furnish men to fish the boats. European shipwrights refused to engage in fishing. This guarantee became increasingly important in the 1890s with the growing demand for fishermen. Between 1891 and 1899 the numbers of fishermen rose from 1,000 to 5,444. In 1902 Alexander Ewen, a leading canner, stated that "the trouble is to get the fish and the people to work. That is the great difficulty." It was a difficulty that was solved by the arrival of Japanese fishermen.

The Japanese, who became the largest ethnic group in Steveston by the turn of the century, had a very different background and work experience than the Chinese fishery workers, yet they were treated in much the same way by the white citizenry. During the first twenty years of the Meiji era in Japan, which began in 1867, there were enormous upheavals in Japanese society, with a rush to follow the Western industrial model. Compulsory mass education started, advances were made in science and medicine, land ownership practices were revised, former systems of class distinction were reorganized, and Japanese citizens were allowed to study and seek business opportunities outside the country.

One result of these upheavals was a growth in population that outstripped the food supply. This pressure led to the colonization of marginal lands in Japan and a limited emigration of labourers to the Americas and Australia, although the Japanese government moved cautiously, having observed the exploitation of Chinese labour in these areas.

The farmers and fishermen of Japan did not prosper under the new order. They were heavily taxed to finance industrial development and little tax money was returned to agricultural concerns. They resented the new compulsory three-year military service for all males aged 21–32 years. Their society had been completely disrupted with no great

Japanese fishermen repairing gillnets, 1913. (F.D. Todd, BCARS)

advantage for them. Many young Japanese, like their European counterparts, were unhappy with their status in their homeland and wanted to improve their prospects. They came to Canada looking for economic opportunities that were nonexistent at home.

A few solitary Japanese had landed on the West Coast in the 1870s, some working in the fishing industry, but immigration in large numbers didn't start until the mid-1880s. Virtually all the newcomers were young, unmarried, adventure-seeking males, who planned to make money overseas and return home with improved prestige and wealth. They were the first Japanese generation to go through the compulsory, universal education system, so they could read newspapers and learn of opportunities outside their villages.

Most of the Japanese who came to Steveston were from Wakayama prefecture in southern Japan. One village of Wakayama was Mio, whose residents made a marginal living from farming and fishing. A carpenter from this village, Gihei Kuno, travelled to Canada in 1887. When he visited Steveston and saw the prosperity associated with its fishery, he was struck by the opportunities available for his fellow villagers. After sending the good news back to Mio, he was joined by a large number of the village's young men who were able to get jobs as fishermen and send money back home. When the fishing was very good, they returned to their village for the winter.

Japanese fishermen posing outside their bunkhouse at Don Island (near Lion Island) above Steveston, 1913.

Once the first group of men gained a foothold in Steveston, they encouraged and assisted others to follow, finding them jobs and providing them with a community of fellow countrymen. Before leaving Japan the young men applied for passports, provided details of their family connections, and passed inspection by the local police. They were sometimes recruited by emigration companies, who received permission from the Japanese Foreign Office to recruit a certain number of men in a certain prefecture. Families of prospective emigrants had to post securities guaranteeing they would fulfil the conditions of their employment. Once in Canada they had to register with the Japanese Consul, and they were expected to return to Japan within three years. These were regulations imposed by the Japanese government with the intention of protecting their citizens from the exploitation suffered by many Chinese emigrant labourers.

The three year regulation was not enforced, and after the emigrant had paid his debt to the emigration company he was free from home country restrictions. The need for cheap labour was widespread in BC, especially in the fishing industry. By 1899 there were 1,955 Japanese in the Fraser River fishery. In the off-season they worked in the coal mines, logging, agriculture or railway construction. In general they worked at heavy, unskilled labouring jobs until they had paid off their debts; the money they acquired after that either went back to Japan or was used to buy farms or businesses.

Before 1907 the Steveston Japanese were virtually all young (average age twenty two years), single men who lived in company bunkhouses during the season, and returned to boarding houses in Vancouver's Japanese area around Powell Street during the winter. Like the Chinese they were organized into "gangs" by Japanese "bosses" who negotiated a contract with a canner. The boss took care of their needs during the fishing season, providing food and sometimes clothing, and services such as mail handling, for a percentage of their wages. They knew little English and didn't associate much with English speakers, their cannery boss acting as translator and intermediary with their employers. Since most of them didn't expect to stay in Canada, they didn't make much effort to acquire the new language, and they preferred the company and familiarity of their fellow Japanese.

In these early years, there was a considerable amount of drinking and gambling by the young Japanese men, especially in the off-season.

The fishermen didn't go to the cannery boss directly when they had claims. The business was handled between the house boss and

the cannery boss. For example, I would say, "This man needs some money, will you lend him some?" The cannery boss would agree to lend some if the house boss became guarantor. The boss controlled twenty or thirty boats. That's how we did things then. You see, in the old days the fishermen didn't have any money, they didn't know English, so even if they wanted to negotiate with the canneries they couldn't, they needed somebody's help. So a Japanese boss talked to the cannery about lending or buying nets and gear for the fishermen. The boss was responsible, very responsible, for them. If a fisherman got into debt, the boss had to pay it off. So he had to be careful to eliminate bad habits like drinking, otherwise he lost money. None of the men had wives then, they were all single. And, as there were no women, they'd get wild. Just a few drinks and they'd start a fight. I saw a lot of fights. Some men killed each other, some were put in gaol. Later, when the brides started coming from Japan, their lives improved. (Asamatsu Murakami, in *Steveston Recollected*)

In 1898 there were only three married Japanese women in Steveston. There were occasional newspaper accounts of Japanese prostitutes being deported, but these were not women from the Steveston area.

The 1890s were a time of anti-Asian antagonism and attempts to restrict their employment and immigration. Between 1901 and 1905 few immigrants arrived from Japan, and the people already in BC consolidated their position in the fishing industry. By the time immigration resumed in 1905, many Steveston Japanese had decided to stay in Canada and had asked their families back

The crew of the *Kinkasan Maru*, a three-masted Japanese sailing ship with a 15-hp auxiliary engine, which loaded salted salmon from the Fraser River. The crew were said to be a "bad bunch when drunk—tearing through the rigging and fighting." The captain asked the community not to supply the men with liquor. (UFAWU)

home to arrange marriages for them. In 1906 these "picture brides" began arriving in Steveston. Some men returned to Japan to find their own wives, others relied on their families to find a suitable spouse. After extensive investigations of background and character, and exchanges of pictures—and if the man liked what he saw—the marriage was registered in Japan, the groom arranged for a passport, and the wife emigrated to Steveston, sometimes immensely disappointed by the difference between the glowing descriptions in letters sent by her prospective husband and the reality she encountered. On the plus side, the women who came to Canada were not under the control of their mothers-in-law as they would have been at home. The "immoral" lifestyles of the young Japanese men disappeared as the Issei (immigrant generation) began to nurture the Nisei (second generation). In the years before World War One, many new families were started. Bunkhouse accommodations were inappropriate for married couples and families, so the canneries built small houses near the dyke for their Japanese fishermen.

These houses were like long barracks partitioned off, very poor, but there was no charge for them. Instead, the cannery bought our fish cheap. No heater in those days, just a [tin stove]—took a lot of wood to keep it going at night. We collected driftwood on the beach, since that was free, and piled it up in front of the house to dry. The house was very cold. It was built of only one layer of wood and we put up papers and so on to insulate it. We had many fires. The stove pipe would rot and develop a hole and the fire would start from there. (Asamatsu Murakami, in *Steveston Recollected*)

No electricity in those days. We used lamps, oil lamps, which we hung up or held by hand. We often had fires because these lamps would tip over and so on. We paid our rent with the fish we caught. We were paid by the piece then and the cannery would deduct five cents from each piece for the rent. The rent would be $50 a year, the water $24. There were good years and bad years, but if we needed it, we could always get a loan from the company. Those houses were cold in the winter.

Japanese cannery housing, 1913. Driftwood was used to heat the houses. (F.D. Todd, BCARS)

Every night we used [egg-shaped hot-water bottles made of tin and wrapped in bags]. Once our water pipe froze and we had to get water in a pail from the Fraser River. I remember we filtered the water through sand and charcoal from the fire before we drank it... (Hideo Kokubo, in *Steveston Recollected*)

There was a road all along the beach side and in the old days, when the tide was very high, or when the snow had mostly melted—well, on the beach side the dykes weren't built perfectly, so if something was a little bit wrong with the dyke, immediately the water would come growling into the road. It was wicked, if there was a tiny gap between the house boards, in it came. Even though there were watchmen at the cannery, there were a lot of unexpected floods. The houses looked like they were standing in a lake. Although they were usually two or three feet above ground-level, the flood sometimes came up to our knees...When the tide went out, the flood was over. Then we repaired them. All the Japanese men lent a hand and the cannery gave us boards...We patched up the holes with clay. In those days we had only a few pieces of furniture, no electric appliances, no valuable things worth worrying about...We built a wooden bathtub which had a tin sheet on the bottom and a wooden rack over that. The water was heated by a fire underneath. We built a ten-foot chimney for it out of clay pipes...The bath was big enough for four people...Of course, not everybody got into the tub at the same time. Some people washed themselves outside while the others were in. You see, in those days we didn't have gas heaters or anything, we had to collect firewood and cut it and dry it. So we had a bath only once a week, otherwise it was too much trouble... (Rokuhei Konishi, in *Steveston Recollected*)

We had to put woollen underwear on the kids, clothes over that, and woollen jackets and stockings. Nowadays you can buy what you want, but in those days I used to make everything by hand—the children's clothing, everything. And things like work shirts—we all borrowed one pattern in turn, adjusting the pattern to fit larger or smaller people. We all did this kind of thing, helping each other out in the neighbourhood... (Umanosuke and Moto Suzuki, in *Steveston Recollected*)

Fishery workers' accommodation. Note the longjohns drying on the railing, and rollers on the windows, for bringing nets inside. (VPL 1395)

Many Japanese women found work in the canneries, doing the same work as the Native women.

When I came [1925]...It was in July, fishing season, so I started work in the cannery almost right away...I packed each can with fish. The work started about June and went to the middle of November. In those days we weren't paid until November. The Chinese contracted for the whole season. They used books of tickets, each book had a 150 tickets. A box contained twenty-four cans. When you finished one box, they'd punch a ticket. We saved these tickets and in the middle of November we went to the Chinese mess house

The Japanese school, 1928. (BSHS)

where they counted the tickets and gave us our money. We used to get $3 for a 150 tickets...In the old days there wasn't enough fish for a whole day's work. Whenever the fish came in, they'd let us know by whistle, day or night. Sometimes we worked one hour a day, sometimes five hours, sometimes ten hours a day. There was a mori house [day care] where the mothers would take turns looking after the kids when we went to work. (Moto Suzuki)

The Steveston Japanese community, like many other Asian communities in early BC, tended to be self-sufficient, with its own network of shops, schools, religious organizations, restaurants and community societies. In Steveston, the Methodist (later United) Church acted as an intermediary between the European and Japanese communities, offering night school classes in English to Japanese.

The poor fishing of the last year obliged us to postpone our plan of a new building there till next year, but the work of the Lord is advancing daily. Our worker, Bro. Nakagawa, has formed a day school for the little folk in the mission like we are doing here in Vancouver, under our new missionary, Miss M. Kudo, and this helps to draw the parents, too. As a result of this work a few children of the school were converted and baptized not long ago.

We keep our night school open on week nights, and it has a powerful influence over the young minds to become followers of the truth. The Japanese fishermen are planning to build a new church for us on the ground where our present building is now standing, as the latter is too small as well as too old. They thus recognize the value of our mission work among them. (*The Missionary Outlook*, May 1904)

Several interesting things have occurred recently. Two women have been baptized and received into the Church on trial. One of them is the bride who came from Japan last spring and was married in our church; the other is the eldest daughter of one of the most prominent Japanese merchants here. Although the father and mother never attend church services, it is a very hopeful sign that they allow their children to come...We visited Steveston a few weeks ago for a couple of days calling on the women and holding a service in the mission building, and realized the great need of a steady worker there. About forty women are there now, but in fishing season the number is greatly increased, and the influences surrounding them at all times are bad. (*Missionary Outlook,* April 1904)

Some early immigrants had converted to Christianity, probably as a consequence of these mission efforts, but once wives and children became a part of Steveston society, Buddhism became entrenched because the Japanese wanted their children raised with traditional values. The first attempts to build a Buddhist temple in Steveston were quashed by white opposition, but a temple was built in 1928. Steveston Buddhism was heavily influenced by its North American setting, with temples resembling churches, and Sunday observances. Important holidays, such as New Year's, were observed and popular entertainments from Japan were recreated in Steveston.

During the New Year's holidays, we often

had a performance of a play or two. It would be an amateur show, but we all enjoyed it. Sometimes, too, we had professional shows from Japan. And there were sumo games. We built our own sumo ring...A few people who did sumo in Japan and liked it taught it here. Judo and Kendo were popular, too. And then there were movies—we didn't have talkies, but we had silent movies accompanied by an orator. I remember the pictures on the screen, and the orator stood beside it and explained it. We had a little hall called the Opera House where these shows were held. (Hideo Kokubo)

The Japanese and Chinese didn't consider themselves "fellow" Asians and never joined forces against the measures being taken to restrict their rights of citizenship. Some Japanese considered themselves superior to all other races, especially the Chinese, and some Chinese resented the fact that the Japanese often underbid them for industry contracts. In addition, from 1894–95 their home countries were at war.

In 1895 the BC legislature extended the clause in the provincial election act barring Chinese from voting to include Japanese, and from 1897 on a series of laws restricted "Orientals" from working at certain jobs. The province requested the federal government to raise the residence requirement for naturalization from three years to ten for Asians. In 1898, BC tried to have the $500 head tax applied to Japanese, and in 1900 the legislature passed a law requiring a language test for all those entering the province. Except for the Elections Act, all these measures were disallowed by the federal government, in most cases because they violated international treaties.

Newspapers in Japan took issue with racial attitudes in BC (and California), and in 1901 the Japanese government suspended virtually all immigration to Canada, although it was probably more concerned about keeping its manpower available for the impending war with Russia. The Japanese, allied with Britain against Russia, had the Canadian public on their side, but their quick work in defeating the Russians and their consequent rise in the ranks of world powers transformed them into the "Yellow Peril" in BC. When immigration resumed again in 1905, the province again started passing anti-Asian laws that were again disallowed by Ottawa. The federal government tried to convince BC that the province had much to gain from commercial relations with Japan, but BC residents of European heritage felt that their interests were being sacrificed to the politics of Empire.

A royal commission examining issues surrounding Asian immigration found many of the problems centred on the fishing industry. Canners

Staff and patients of the Japanese fishermen's hospital, c. 1897. Dr. R.R. Robinson is at centre, Miss Simmons (a nurse) is seated, Miss Abrams (a nurse) is standing. The surgeon at the hospital was Dr. Oishi. (CVA)

were very enthusiastic about their Japanese employees, who not only provided cheap labour but were the best fishermen, consistently outfishing all other ethnic groups. The Japanese had also come to control the fishboat-building industry. Since they lived in conditions whites considered unacceptable and worked for less, the white working class felt threatened.

In 1900 the white fishermen's union invited Japanese fishermen to join them in addressing matters of mutual concern, but the Japanese had already formed their own protective association. The Japanese Fishermen's Benevolent Society, or Dantai, dealt with everything related to the welfare of Japanese fishermen, and nearly every Japanese worker belonged. One matter addressed by the Dantai was the health of its members and their community. The housing and water supply available to the Japanese in the 1890s appalled a visiting Japanese American dentist from Portland (see Chapter 4); typhus and yellow fever were often blamed on the unsanitary conditions that prevailed in the summer. The dentist instigated the building of a chapel-hospital on land owned by the Phoenix Cannery, to be administered by the Dantai. A doctor and nurses were brought out from Japan. Dr. Large, of the Methodist mission at Bella Bella, helped out in the summer when he accompanied the Natives to Steveston.

> **Japanese Hospital, Steveston, B.C. (These Hospitals, although not under the control of the Methodist Church, are valuable adjuncts to the work). The Japanese Hospital is supported by the Japanese. Its success is largely due to the Christian Japanese in connection with our work in Vancouver. Dr. Large, of Bella Bella Hospital, while at Steveston for the fishing season, acted as visiting physician. The growth of the work and the demands of the thousands who gather at Steveston upon the "Doctor" has made it necessary to have a resident superintendent for the Japanese Hospital. The Christian Japanese from Vancouver act as nurses. (*The Missionary Outlook,* August 1904)**

The Japanese also started a co-op where members could buy food and clothing for better prices than elsewhere. During the fishing season, fishermen were obligated by contracts to sell their catch to canneries, but out of season they could sell anywhere. Some fishermen started a business shipping dried fish to Japan, much of which was sold through the co-op.

The earliest Japanese immigrants became naturalized Canadians as soon as possible because it was necessary to be a British subject to obtain a fishing licence. Most intended to return to Japan, so they did not feel themselves particularly affected by the provincial law denying them the franchise. But because their names were not on the provincial voters' list, the Japanese were also excluded from federal voting lists (made up from the provincial lists) and from municipal and school board elections. Although they did have to pay taxes or be available for conscription, they could not be elected to the provincial legislature, municipal or school board office, or serve on a jury. They were also barred from occupations whose qualifications included being on the provincial voting list, such as public works, holding a hand-logging licence, forestry work, law, post office employment, police work and many others. The franchise became an issue when Japanese people began seeing Canada as their permanent home, where they would be raising families and aspiring to jobs other than those in the fishery. The law was challenged in Vancouver by Tomekichi Homma in 1900. He was successful in the Supreme Courts of British Columbia and Canada, but lost when the matter was referred to the Privy Council in London, on the grounds that the British North America Act gave provinces exclusive jurisdiction over civil rights, such as the right to vote.

Many British Columbians also wanted to exclude the Japanese from becoming British subjects, claiming the process was being abused in order to get fishing licences. The accusations were in most cases too difficult to prove in court, but there was probably widespread abuse of the practice, especially by labour contractors who gave their non-English speaking crews "papers" which they understood only as something that enabled them to fish. These tensions set the stage for bitter disputes when labour unrest manifested itself in the Fraser fishery.

When Japanese immigration resumed after the Russo-Japanese War, the province's economy was in a temporary recession. The fears of Japan becoming a powerful nation and of Japanese workers undermining white labourers' wages, and rumours of Japan preparing to take over the Panama Canal all came to a head in 1907. The steamer *Kumeric* landed in Vancouver with over a thousand Japanese on board, eight hundred of whom were taken to spend the night in a cannery warehouse in Steveston. When more boats followed, there were fears that an invasion had begun. In September 1907, an anti-Asian demonstration turned into a riot that raged through Vancouver's Chinatown and "Little Tokyo," inspiring editorials and discussion around the world. The Japanese government agreed to restrict the number of passports issued to male labourers to four hundred a year. In the 1908 federal election, Japanese immigration was a major issue, and a "White Canada" campaign was very effective for the Conservatives who, despite losing nationally, picked up several seats in BC. The exclusion issue figured prominently in provincial elections as well.

The restrictions on male immigrants (but not their brides) cooled tensions somewhat, and the

pre-World War One years were relatively peaceful, though the province was still attempting to implement language tests and other exclusionary measures. The war distracted attention from the immigration issue. Fears that the relatively unprotected West Coast was vulnerable to attack from Germany's China Squadron were relieved when two Japanese men-of-war entered Vancouver harbour in compliance with the Anglo-Japanese alliance. The ships patrolled the coast throughout the war and probably offered some measure of protection to the CP *Empress* ships. Some people suspected the Japanese in Steveston might be agents of Japanese imperial ambitions, but the premier didn't want to provoke racial unrest during the war, especially when there was a shortage of labour. Some Japanese wanted to enlist and trained themselves at their own expense. Eventually 196 went overseas; of these 53 died and only 10 returned unwounded. In France, Japanese Canadians were allowed to vote in the 1917 federal election, and on their return to BC they pressed for the franchise and for equal fishing rights.

The postwar years were not good ones for Steveston's Japanese fishermen. There was an economic recession, jobs were scarce, and there was a feeling among white people that the Japanese were monopolizing the fishing industry because they would accept low wages. They had consolidated themselves in the industry during the war; high prices and labour shortages had benefitted them economically, and by 1919 they controlled about half the fishing licences. The 1913 Hell's Gate slide had greatly reduced fish stocks, and white fishermen had a hard time competing with the highly skilled Japanese. There was a movement to get them out of the fishery to make more jobs for whites. The Fraser River Fishermen's Protective Association, founded in 1914 to organize the anti-Japanese elements, was revived and started campaigning for anti-Japanese legislation. The federal government bowed to this pressure and made plans to gradually reduce the number of licences issued to "other than white resident British subjects and Canadian Indians." In 1923 Japanese licences were reduced by 40 percent and were tied to specific area of the coast. Other fishermen were still permitted to move up and down the coast as fishing conditions changed.

Japanese children at play in Steveston, 1930s.
(James Crookall, CVA 260-611)

The government's actions did not go unchallenged.

During the First World War there was no movement to chase the Japanese out of fishing. But after 1919 the government decided not to allow the number of Japanese fishermen to increase. Then in 1922–23, they began to cut down on the number of licences, but only for

The Canadian Navy confiscates Japanese fish boats, December 1941. Note the shadow on the boat. (NAC 170513)

the Japanese. So Japanese fishermen decided they had to do something about it and they went to talk to the government, with no result...Then a committee decided in 1926 to get rid of the Japanese in ten years, which meant there would be no Japanese fishermen in BC by 1937. At that time, in Steveston alone, there were more than five hundred fishermen. By the end of this court case I'm telling you about there were only two hundred left. About 1926, a lot of Japanese agreed that the only thing to do was to file a case in court [they had gone to the Japanese Consul]. At that time the Japanese didn't have the vote, but the Consul said Japanese fishermen were responsible for the problem because they hadn't cooperated with white fishermen. Then a new consul came in after him and in 1926 and '27 we filed a suit in the Supreme Court. One clause of the Fisheries Act said that the minister may revoke a licence and our lawyer took this 'may' to mean that this limited the Minister's decision-making power. And our lawyer also said it was against the Canadian constitution for the minister to take away the licences of Japanese Canadians [naturalized]. That decision in our favour came in May of 1928. But after that, Parliament eliminated the Act and they filed their case with the Privy Council in London. The case was tried in England in 1929. The lawyer for the Japanese side was Sir John Simon, who was one of the best in London at the time. We were able to hire him because the Japanese Embassy in London helped us. John Simon said it was against international mutual agreement and also against the Anglo-Japanese accord to discriminate between people in one country who were all supposed to be British subjects. So we won the case. But in BC only about half of the fishermen remained by the time this court case was over. (Rintaro Hayashi, in *Steveston Recollected*)

As the Nisei came of age, they found many areas of employment were closed to them, regardless of the education and experience they had acquired. Despite the restrictions imposed on the fishery, many continued to find employment there, though often they had to leave Steveston to work in the northern coastal areas. In the 1930s there were renewed but still unsuccessful attempts to acquire voting rights. Then, in the late 1930s, anti-Japanese feeling renewed as a result of Japan's involvement in Manchuria and China. Although RCMP surveillance of the Steveston community found no evidence of subversion linked to these events, there was a widespread belief among the white public that BC was also a potential target of Japanese aggression. Prime Minister W.L. Mackenzie King set up a standing committee on Orientals in BC to assure people that the government was keeping an eye on the situation and to defuse any potentially violent agitations that might provoke Japan to declare war on Canada.

With the attack on Pearl Harbor there was no way to placate the fears. Early proposals to remove enemy aliens (Japanese citizens, about one quarter of the Japanese in Canada) and adult Japanese males (regardless of their citizenship) were rejected, but on February 27, 1942, the federal government gave in to the demands of some BC politicians and ordered the removal of all persons of Japanese ancestry from the coast. Most were sent to ghost towns in isolated areas in the province's interior, while others laboured on sugar beet farms in Alberta and Manitoba. All the real and personal property they left behind was confiscated; their fishing boats were immediately sold and put back to work. The government intended to hold the internees' goods in protective custody through the Custodian of Enemy Property, but in 1943 they were sold for a fraction of their value. The meagre proceeds, less the costs incurred in storing and selling the goods, were held for the owners who could use them to support themselves during the rest of the war.

During and after the war the third generation, the Sansei, were born. When their parents were released from the detention camps, only about a third of them returned to the coast. A great many moved to the East, especially Toronto, and several thousand went to Japan. Those who came back to Steveston had lost both their property and their community structures, and they increasingly integrated into the larger Canadian society. Japanese Canadians won the right to vote in federal elections in 1948 and in provincial elections in 1949. The Sansei were able to enter occupations in many sectors of the labour force that had been closed to their parents, and there was increased intermarriage with whites, a phenomenon almost unknown in the Issei and Nisei generations. Today, their links with their heritage are retained in a continued connection with the fishery as gillnet fishermen, in the Japanese language school and Martial Arts Centre in Steveston, and in the preservation of cultural activities such as folk dancing, gardening and floral arrangement in the Japanese style, and other art forms. In the 1970s, a small group of Japanese Canadians began a movement to seek redress for the internment years. A negotiated settlement with the government of Canada, signed in 1988, provided an official pardon, financial compensation to individuals, an offer of citizenship to Japanese Canadians who had been exiled to Japan, and funds to help rebuild communities that were destroyed.

Confiscated Japanese-owned boats at Scottish Canadian cannery. Steveston, 1942. Note Canadian naval personnel on boardwalk. (NAC)

Living quarters for Japanese workers, early 1900s. (VPL 27244)

A gillnetter being taken into Kishi shipyard, 1970s. The building later burned to the ground. (DS)

A recent photo of Mr. Jim Kishi, propietor of Kishi shipyards. (BSHS)

CELAINE
NEW WESTMINSTER
B.C

When the first canneries were built, there were no laws for limited liability or even for registered partnerships. Canneries were sole proprietorships, associations, partnerships or companies, and were often started by men with gold rush or related experience who were looking for a safer place to invest their resources. These men arranged with a fiscal agent or commission house, usually in Victoria or San Francisco, to import canning machinery and supplies from Britain, and to ship the product to market, again usually Britain. Because it could take up to eighteen months to fill an order, with no guarantee that a shipment would not be lost at sea, cannerymen operated on loans advanced by commission houses. It was a risky business for the canner, though probably a very profitable one for the agent. Some men became wealthy, but success depended on fluctuating British market conditions, and canneries frequently changed ownership—and names. Before the arrival of eastern banks in British Columbia, canners depended on these arrangements since mail service to eastern financial centres was too slow. Once the banks opened, however, individual cannerymen exercised more control over their company's operations, purchasing materials at lower prices whenever possible, and introducing new plant operating methods.

For example, some Steveston canners saw an opportunity to reduce costs by having their salmon loaded directly onto sailing ships rather than trans-shipped them to Victoria. When the China tea clipper *Titania*, at that time owned by the

Steveston canneries, looking upriver, c. 1913. Note the powered gill-net vessels. (BCARS 84110)

Hudson's Bay Company, had extra space available on an 1889 voyage that included stops at Victoria and Vancouver, the space was filled with cases of Fraser River salmon, loaded directly at the Britannia cannery wharf. The *Titania*'s September arrival in Steveston was the first time a Europe-bound vessel had docked there. Not only the canners benefitted; ship owners were pleased to have a fresh water port in which to rid their ships' hulls of barnacles. Over the next several years many more sailing ships, with their exotic crews, moored in Steveston's harbour. Then the *von Galen* nearly tore away the mooring piles in front of a cannery during a strong wind. Extraordinary emergency measures prevented a disaster, but marine insurance rates on sailing vessels entering the river shot up to a prohibitive rate. Shortly afterwards, Fraser River canners started depending on the railway to ship their product.

When legislation in the 1890s created the legal format of the limited company, there was a boom in the cannery industry. Businessmen entered the field and the number of canneries and the total output of canned salmon skyrocketed. The first of these companies, British Columbia Canning Company, was formed in 1889; by 1891 limited companies such as Anglo British Columbia (ABC) and the Victoria Canning Company controlled over 60 percent of the Fraser River's pack. In 1902 the British Columbia Packers Association was formed. This limited company alone absorbed twenty-nine of the existing canneries and twenty-two existing firms, including the Victoria Canning Company. In its first year of operation, BC Packers controlled over 50 percent of the Fraser pack. The new corporations radically decreased the importance of local capital, especially that of commission agents. BC Packers was backed by a consortium of eastern Canadian financial interests and ABC was backed in the United Kingdom. Altogether, the rise of the corporate structure profoundly changed the nature of the fishing industry, moving it out of the hands of frontier entrepreneurs and into as stable a position as one can find in the fishing business.

The technological developments in the first twenty-five years of the fishing industry were intended to save labour costs. Although the average production remained constant, the size of an average cannery crew decreased from 120–150 to 85. BC Packers quickly started to consolidate its plants in order to further decrease cannery labour, the price of fish, insurance costs and the operating costs of its canneries. By 1905 the company had reduced the number of its operating canneries on the Fraser from twenty-nine to fifteen, with four operating more than one canning line. Equipment for these additional lines was obtained mainly from abandoned canneries. It was thought that centralizing plants would decrease labour needs. Plants with two or more lines did not have to stop and change cappers when a change was made in the style of the can; as a result, they were more productive. Abandoned plants were either sold or used as storage warehouses, alleviating the problem of finding space for processed cans awaiting shipment. The reduction in the number of plants greatly reduced insurance costs, which were high due to the risk of fire, especially in the Steveston area where canneries were closely crowded together.

Another innovation on the Fraser was the application of electricity as a source of power and light. In the first decade of the twentieth century, various machines were developed for mechanized can filling, weighing and salting. The modern canning line was finally achieved in 1913 with the introduction of the sanitary can and the double seamer, which eliminated the soldering of can lids.

The results of using the sanitary can were phenomenal, both in the volume of output and in the quality. The percentage of leaks and "do-overs" was reduced to a minimum. The Chinese experts who formerly operated solder machines, mended leaks, and did the venting and stopping were no longer needed. The sealing process required one-third the labour of soldering. The first cooking, with its subsequent venting and stopping, was no longer required to prevent the cans from bursting in the retorts.

New warehouse at BC Packers' Imperial plant, 1950. (BCP)

In the fishing sector, the gasoline motor revolutionized the gillnet fishery and made the purse seine fishery economically feasible. The introduction of gasoline engines to the Fraser River gillnet fleet started at the turn of the century when Easthope Brothers, a machine shop in Steveston, produced the engine. Pioneers on the Fraser set the date of the first gasoline-powered salmon vessels between 1902 and 1905. Very few boats had engines before 1906, however, and 1907 is the first year gasoline boats are listed in the BC Fisheries Report, so they could not have played a major part in the fishery before then.

The gasoline engine was adopted slowly because fishermen were sceptical of its utility and staying quality. Early engines were light and frequently broke down. At first, most engine parts were available only from the East, which meant long delays for repairs. With the development of stronger engines and the establishment of local repair shops, such as Schaake Machine Company, Easthope and Sons, Canadian Fairbanks, Letson and Burpee Ltd. and Henry Darling—all in business by 1907—the use of gasoline equipment rapidly increased. By 1910 gasoline-powered gillnetters comprised 50 percent of the Fraser fleet, and by 1913 over 80 percent were mechanized. Even the oar and sail vessels benefitted from this new technology. The powered boats could tow several skiffs in a row and drop them off along the line.

The only design change needed in the traditional Columbia River gillnet boat to adapt it for gasoline was to alter the stern to take a propeller. A small house in the bow replaced the pup tent made from the sail. These engine-driven Fraser River gillnetters were commonly referred to as the "mosquito fleet." Early engines were of the three to five horsepower variety and made six to seven knots. Most of these engines were two cycle and were produced by Easthope and Sons, Cowie, Vivian, Adams, Toronto Junction, Buffalo, Gray, Frisby, Palmer and Hyannis.

The rapid adoption of engines after 1907 was a result of competition for the resource. When one gillnetter successfully used an engine, competition forced other fishermen to follow suit. Mechanized gillnetters had a distinct advantage over oar and sail- powered vessels. They could make more sets because they could move more quickly upriver to start a new drift, and they could work farther offshore. Fishing time was increased because less time was required to travel to and from the grounds, and motorized vessels could fish in rougher weather. Engines also eliminated the backbreaking labour of manning the oars:

Tide, wind and sea which would interfere with old style fishing seldom troubles the modern motor boat fisherman.

With the motor doing the hard work for him, he can twirl the wheel or keep a hand on the tiller and rest up after the labour of fishing...The work of fishing is hard enough without adding to it the strenuous labour of sail handling in windy weather and winter gales. (*Canadian Fisherman,* February 1914)

The gasoline motor was probably the most important factor in the development of the purse seine fishery. The purse seine used the principle of encirclement and was used mainly for the schooling species of salmon—chums and pinks. Unlike sockeye, red springs and cohoes, which travel at various depths, chums and pink salmon school at or near the surface, making encirclement, or seining, practical. The beginning of the seine fishery marked a new concentration by Fraser River canners on the previously ignored "lesser" species. In the 1911 season all five varieties of salmon were used for canning to a greater extent than ever before.

The early purse seines were a webbing of tarred cotton with a lead line and cork, or float line, measuring 175 to 225 fathoms in length. Brass rings were hung at regular intervals along the lead line by means of "bridles." A purse line was then passed through these rings. Salmon purse seining was introduced to Puget Sound in the 1880s. It was conducted from two large flat-bottomed scows equipped with little more than a hand-powered purse line winch. At the beginning of each season a steam tender towed the scows to the grounds, where a large skiff manned by eight oarsman pulled the scows from place to place to set the net. The tenderboat was a vital part of this early purse seining technology. It towed the gear to and from the grounds, and transported the haul to the canneries. This early method of propulsion limited purse seining to a few miles around each fish camp, but it was an improvement in terms of mobility over the beach seine system. The purse seine was also far more efficient at trapping fish than the beach seine.

As in the case of the gillnet fleet, the introduction of the gasoline engine rapidly increased the mobility of the seine fleet. Powered seiners used a different fishing technique:

The scows were discarded and the net moved from the skiff to the powered vessel. The end of the net was now made fast to the skiff which acted as a buoy and the seiner ran a circle setting the net and returning to the skiff. Both ends of the net were then brought aboard, the net pursed and finally hauled or 'dried up' so the fish could be removed. ("History of Western Seining," *National Fisherman,* 1971)

The typical early seiner was a small open boat decked forward with a small house over the engine. It had no crew quarters. Hulls were very beamy to take heavy cargoes and the strain of pulling in the nets; the beam was 25 percent of the length. Engines were only five to twelve horsepower because they took up the same space as a modern 100-horsepower engine. The net was stored aft on a "table" or platform. Although the earliest seiners had only hand-powered purse winches, these were

Louie Korens in BC Packers' Sunnyside Cannery boat #26, fishing off Steveston, 1940s. This boat was brought down to Steveston from the Skeena and had a drum added for the Fraser fishery. Note the Steveston fuel barges in the background. (NAC)

soon mechanized to eliminate the heavy work. The purse winch was connected to the main engine with a line shaft. Mechanized seiners could make twice the number of sets in a day.

Steveston's seine fleet developed rapidly after 1911 to exploit the Swiftsure Bank in the Strait of Juan de Fuca. Between the salmon seasons of 1911 and 1912, the number of seiners in this fishery rose from twenty-two to one hundred. They were open water boats known as deep-sea seine boats. They differed from the early seiners in that they were fully decked. The Swiftsure Bank seine fishery was another attempt by the Fraser River canning industry to intercept the salmon runs before they got to the Americans.

The internal combustion engine had a two-fold effect on the steam tenderboat fleet. Mechanization of the gillnetter decreased the steam tender's importance since tenders no longer transported the bulk of the gillnet fleet's catch to the canneries. On the river, small gasoline-powered gillnet collectors—actually gillnet boats with an open or closed hold filling the whole stern aft of the house—began to replace the tug and scow method of transporting fish. With the move to salmon seining in the Strait of Juan de Fuca, tenders provided an essential transportation link to the Fraser River canneries. Transportation of salmon from these new grounds by the tug and scow system, however, was very expensive. As the distance increased, perishability of the catch became a matter of concern to the canners; a new type of tender, the diesel-powered packer, was introduced. This vessel had its own holds with pen and shelf boards to prevent crushing the salmon. Because the hold was covered, ice could be used economically, which was not possible in an open scow. Packers were rare before diesel motors because existing gasoline engines were uneconomical for such large vessels. Diesel did not play a major role in the fishery until after 1913.

The new ways of organizing the fishing business made it possible to bring more money into the industry and thereby automate and mechanize in ways undreamed of when the first tins of BC salmon were produced. The rapidly changing technology and the other pressures of a big, modern business were not without social consequences, many of which were most vividly felt in the white community.

In Steveston's early years the farming community was almost exclusively of European descent, mostly from the British Isles. They also participated in the fishing business once it was underway, joined by Europeans from many other cultures, especially those with a fishing tradition. The "whites" receiving fishing licences in 1912 were categorized further as Canadians, Scandinavians, British, Austrians, Greeks, Finlanders, Italians, Spaniards, Germans, French, Russians and nine "indescribables." Many of the Old Country techniques had to be forsaken in the New World, however, and like every other ethnic group involved in the fishery, Europeans had to learn the unique characteristics of Fraser River fishing.

All the Fraser's cannery owners were white men with the exception of John Deas, a black American and an experienced tinsmith who had practised his trade in the southern US and in California before emigrating to British Columbia. In the early years most of the whites worked as fishermen. Those working in the canneries were generally in supervisory or clerical positions. Some white women worked on the canning line, mainly lacquering cans. Later, especially in times of labour shortages, white women could be found on other parts of the canning line and in various types of clerical work.

Steveston's first cannery, Phoenix, was built by Marshall English.

He was a tall and very erect Virginian of the old southern type—a man of fine gentle-

Opposite Page:
Gillnet full of salmon. (BCP)

Below:
The Steveston waterfront, looking upriver.

manly manners. He was splendid company. He was a good business man, but was unable to refrain from having a good time while the money lasted. He could drink like a fish but was always able to carry himself. Sometimes, asking to be excused, he would step aside from others he was walking with, and would spill the surplus in the ditch at the roadside. He then would rejoin his companions and resume the conversation without showing any discomfort from his experience. Shortly thereafter, he might even join others for another drink. (T. Ellis Ladner)

Alexander Ewen had an interest in Canadian Pacific cannery and had a somewhat different reputation: "Alex Ewen was a dour Scot and extremely canny. I don't recall anyone saying he had heard Alex Ewen laugh. He did, however, have a temper. It was generally shown only as evident impatience, but, on occasion, it was more forcefully displayed." Other canners, such as J.H. Todd, owner of Beaver Cannery, "was not of the cannery men who fraternized" (T. Ellis Ladner).

In addition to the financial challenges, the early canners dealt with an often volatile workforce. Labour shortages were a constant problem, and when they started using Chinese labour, other problems could arise:

He [Marshall English] had a white employee in the cannery who was a domineering bully whose special delight was to play mean pranks on the inoffensive Chinese workmen that laboured under him. He picked especially on one oriental, a harmless quiet man who plodded steadily on at his task under provocations that would have aroused the anger of almost anyone else. But even the most timid of persons has a limit to his stock of endurance and one day the ribbing this Chinaman was subjected to was exceptionally severe and because he bore this he was being pushed around even more roughly than ever. Finally the victim protested angrily. Thereupon the white bully knocked the Chinaman down and the latter, striking a table as he fell, arose with a nasty cut on the top of his head, with blood flowing freely down the side of his face. Immediately the other Chinamen present, who had witnessed the assault, set up a "ki yi" and made for the offender. Like all bullies he became frightened when menaced himself and called on the cannery foreman for help. The latter on coming to the rescue struck down one of the orientals with a can tray he had seized as a weapon whereupon every Chinaman in the place, most of them armed with large sized fish knives, joined in the fray and the whites, badly scared, rushed for the door and scurried through to safety with the Chinamen in hot and angry pursuit behind them. Mr. English was on the point of entering the cannery when the white men in panic terror rushed past him on their way out. He realized at once something untoward had happened but had no time to enquire as to the cause for he saw, and heard, the armed and angry Chinamen rushing toward him. Quickly he stepped inside the building and as quickly shot the bolt that locked the door and then turning he, with arms folded across his breast, faced the excited mob that was seeking vengeance. His cool courage momentarily cowed the angry men and when one of the few who could talk English recovered sufficiently to tell him to get out of the way so they could pass he sternly refused and, through their Chinese foreman, ordered them all to stop their foolishness and return to their work. Thus alone and defenceless he faced twenty or more maddened men, bent upon attack, and they, awed by his bravery and cooled down by the delay, silently submitted and resumed their work. When order had been restored he examined into the cause of the disturbance and immediately discharged both the bully and the foreman; the first for his treatment of the Chinamen; the latter for taking the offender's side without first ascertaining who was in the wrong. From then on there never were any labour troubles in the cannery of English and Co. (Henry Doyle, *Rise and Decline of the Pacific Salmon Fisheries*)

White men of lesser means worked as fishermen. In the industry's first days, when white labour was scarce, Natives dominated the fishing activity. By the mid-1890s, however, Natives held only 43 percent of the net licences on the Fraser, while whites had 37 percent and Japanese 20 percent. By the 1901–1904 period, white fishermen had 44 percent of the net licences (as did the Japanese; Natives held only 12 percent).

Before 1892 most of the white fishermen were hired by a cannery, used the cannery's boats and nets, worked under a company licence, and received a wage. Eventually canners shifted away from a daily wage to paying by the piece. By the mid-1890s many fishermen owned their own boats and gear, held their own licences, and made contracts with the canneries to supply fish to them. There were numerous complaints to the fisheries department about the practice of limiting the number of licences issued to fishermen, and in 1892 the federal government bowed to local pressure and lifted the restrictions. This led to new problems:

> **This overcrowding also decreases the individual catch and forces the fishermen to demand more for their fish than formerly, and receive less remuneration for their work, both the canner and the fisherman losing thereby. It has created serious irritation between the white fisherman and Japanese, the former complaining that they are forced out of an industry where unskilled labour is employed, who work at very low wages, have no family to support and send or take most of their earnings out of the country. As long as the fishing is profitable to the fishermen so long will the white man be willing to engage in it. As the margin of profit grows less, they will drop out... (*Report of the Royal Commission on Chinese and Japanese Immigration*, 1902)**

Not all whites worked as fishermen or on the cannery lines. Others worked in the canneries as machinists, setting up and maintaining the canning line before and during the season, as foremen and overseers, and as clerks. They were generally unwilling to work at the low wages paid to Chinese workers. During one Royal Commision, a canner described the duties of various workers:

> **Q. How many hands are employed in a cannery?**
> **A. Well, in the neighborhood of 100, all told—that is, in a cannery with a capacity of 15,000 cases, because there are only a few days when all**

Housing for cannery workers. (BCARS 82291)

can be working fully.
Q. Of those 100 hands, how many are foremen, generally?
A. I have been a foreman—there would be a foreman of the Chinese—he would be a Chinaman.
Q. Yes; but how many white men in specified positions?
A. Well, then there is the foreman of cooking; then there are firemen, and several other assistants.
Q. The firemen would be white men, and all the rest Chinamen?
A. No; there would be a white inspecting the filling, and white men would be in charge of the retorts, timing and keeping the proper temperature; then watchmen, etc., about eight white men.
Q. Would that be a fair average in other canneries?
A. Yes; there might be days when they would bring in one or two others.
Q. And the rest?
A. Klootchmen and Chinamen—the greater number Chinamen...
Q. Then the white labour employed in a cannery, turning out 15,000 cases, are some six or eight white men?
A. Yes; but I may say it would not pay any white to do the work the Chinamen do for the pay, or anything like what the canneries would be willing to pay. (BC Fisheries Commission, 1892)

The whites had the highest salary levels, most of the management jobs, and the best seasonal housing conditions.

Lastly lets take a look at the food supplies. As for our own table we had nothing to complain about. We had good Chinese cooks who made freely, pies cakes cookies &c. We were much better off than logging camps in that we had unlimited fresh vegetable, meats and fruit in season, eggs, milk butter &c. (Edwin DeBeck)

Some of the management and fishermen lived on the cannery sites year-round. The fishermen's accommodation was either single family detached homes, duplexes or triplexes, modest but comfortable. They were generally painted white and trimmed with the company colours, usually red or green. The houses occupied by management and their families were more elaborate:

[the house] was given to George 'cause he was manager of the Britannia Machine Shop. Different managers put on rooms, added rooms to it. It was built by Mark (Marshal) English, the man who first came and built the first cannery on the Fraser River, and he built a small bungalow for his family and he brought them from Virginia. And then these managers from the canneries over the years, built three very large rooms, one beautiful living room with a twelve-foot ceiling, and a large bedroom and a large kitchen to it. So we had...three lovely bedrooms and a very large kitchen and a large living room and back rooms too, for storage, for canning. I used to stand on a very large trellis and paint the twelve-foot ceiling too, it was a v-joint ceiling. Then we panelled the walls all in English panelling, and put a plate rail all along the top and then wall paper from the plate rail to the ceiling. We had a beautiful room with windows with nice frilled curtains. It was very happy, very pretty.

Mark English put the orchard in when his family arrived. By the time we were there, it had begun to produce fruit and had apples, and plums, and pears...And a nice stream running through there beside the house and at the back. We had a beautiful garden belonging to the house with ivy fence, and nice rockery in the corner with acacia trees. Some of the managers had come from California, they'd brought roots of acacia trees from California and planted them. (Dorothy Mackenrot, in *Steamboxes, Boardwalks, Belts and Ways*)

By contrast the fishermen's houses, or "shacks," were more modest affairs:

What they used to do is bring out about six or eight 2 x 4s and they'd build these houses and they'd go from house to house with the 2 x 4s, so there's no frame [in the walls]. These is no 2 x 4s in this thing. Originally when they used these 1 x 12s, they'd put one [2 x 4] at the top and one in the middle and one on the bottom. Then they'd take these 1 x 12s and nail them on and they had battens [or 1 x 3s]...every house had to have a lean-to on it for some reason. (Gerry Miller, in *Steamboxes, Boardwalks, Belts and Ways*)

When the managerial class finished a week's work they often unwound with entertainment in their homes:

> **We had a lot of friends you know, great many friends, and they were good [bridge] players too. We had as many as 6 [bridge] tables. We had a lot of fun. And we'd have dancing, and music. And I could play the piano a bit at that time. But then my mother, when she came to live with us, she was a wonderful player, and she could play the piano for them too. And we had the radios and a big room to dance in, we'd just roll the carpets up and have a dance, and everybody danced.**
>
> **We had one of the sheds up there, that they used for the nets, they didn't need it any more so we turned it into a badminton court, and we played badminton there. Then George and I used to play badminton in Richmond too, at the badminton courts. We were great on badminton.**
>
> **When we first went there, you know, we could go swimming. We'd go right down from our house into the river and the water was clean and beautiful and clear. (Dorothy Mackenrot)**

Many members of the white working class arriving in BC from Britain and the US were familiar with and perhaps involved in the socialist and union movements. Considering all the different types of occupations in the fishery, the complications of racial and gender attitudes, and the competitive rather than co-operative nature of the fishery, it is amazing that any common ground was found on which to stand as a union. It wasn't easy and it took a very long time to organize the present-day United Fishermen and Allied Workers Union (UFAWU).

The fisheries labour force attracted white, Native, Chinese and Japanese workers, and included a high percentage of women and some children. These workers weren't necessarily pulling for common goals. Some white fishermen's associations excluded "Orientals" while others encouraged Asian membership. One faction of Japanese Canadian fishermen was a protective association against whites, while another Japanese labour union worked for closer ties between whites and Japanese. One faction of Natives was pro-Japanese; another was anti-Japanese. The truth about the fishery is that while it sometimes unites its workers—for instance by mutual dependence in the face of dangerous working conditions—it also separates workers, as each small fishing unit works independently on the fishing grounds and fishermen are in port together for only a short time. Most of the boats are individually owned and compete with each other for a share of the resource. In the early years of this century, even more than today, racism and sexism were part of the structure of West Coast society, and thus part of the structure of Steveston's fishery.

Competition for the resource resulted in conflict among the various fishing gear types (seine, gillnet, troll, etc.), and these gear conflicts crossed racial, gender and political lines. The dozens of fishery unions and associations in the early years existed mainly because of equipment differences.

There were work stoppages over fish prices and working conditions almost from the start of the industry, but the most famous incident occurred in 1900. Between 1890 and 1900 the number of canneries on the Fraser increased dramatically, resulting in intense competition among canners. By 1900, however, they were united in their desire to prepare for the coming peak year in 1901 by arranging for a stable packing season. To this end they created the Fraser River Canners Association, to regulate the number of fishing boats and to set a standard price for fish. The fishermen responded by organizing a union with locals in Vancouver and New Westminster. When the two sides confronted each other in July 1900, the major issue was the price of fish. Canners wanted to regulate the number of boats and the price of fish, a move the union saw as heralding a lower price for the 1900 "off" year and an even bigger drop in 1901. Against the advice of union leaders, who recommended not going over twenty cents, the fishermen, meeting on a field outside the Steveston Opera House, voted to demand twenty-five cents per fish. The Japanese and Native fishermen had been invited to join the union but both had declined, the Natives remaining with their tribal organizations and the Japanese supporting their own Japanese Fishermen's Benevolent Society organized two months earlier. Both Japanese and Native fishermen initially supported the strike, but as the season wore on the Japanese grew eager for a speedy settlement. More than the other groups, they were dependent on the canners for their seasonal food and shelter; the salmon catch might be their only income for the year.

The canners' offer was twenty cents, and demonstrations against this were peaceful. At the end of the second week of the strike, the canners attempted to negotiate via a representative of the Department of Labour, who suggested an offer of twenty cents, to be reduced to fifteen cents if the run was heavy. The canners would also agreed to take all fish caught and not impose any boat limits. Both the union and the Japanese rejected the offer. The following week the fishermen asked for a price of twenty cents throughout the entire season, the right for independent fishermen to deliver to any cannery, and equal boat limits, if imposed, for both cannery and independent fishermen. The canners rejected this and reduced their offer to fifteen cents across the board.

The canners sent strikebreakers out to fish, but they were brought back to shore and assaulted by the fishermen. Special constables were sent to Steveston and a union leader was arrested. On July 22 the canners made a final offer of twenty cents for the first six hundred fish and fifteen cents for any additional ones. The union rejected this offer, but the Japanese accepted and agreed to return to work the next day. Most of the Japanese fishermen could not speak English and had to rely on their boss contractor for information about the strike. Like the Chinese contractors, these men had arrangements with the canners and were paid bonuses for supplying fishermen.

The canners thought the fishermen would retaliate against the Japanese, and asked for military protection. Both the provincial and federal governments refused. Canners could call out the militia, however, with the signatures of three justices of the peace, a relatively easy task as one JP was a cannery owner and another a cannery foreman. On July 23 the Duke of Connaught's Own Rifles arrived in Steveston from Vancouver on the steamer *Comox* and set up camp in front of the Gulf of Georgia cannery. Union fishermen dubbed them the "Sockeye Fusiliers." On July 24 most of the residents went fishing but the whites and Natives, particularly the Native women who worked on the canning line, continued the strike, forcing the canners back to the table where both sides finally agreed on a price of nineteen cents per fish on July 30. The militia went home, the only casualty being a hand cut on a bayonet.

Early yesterday morning the first and only collision that has occurred between the civil and military forces was reported from the

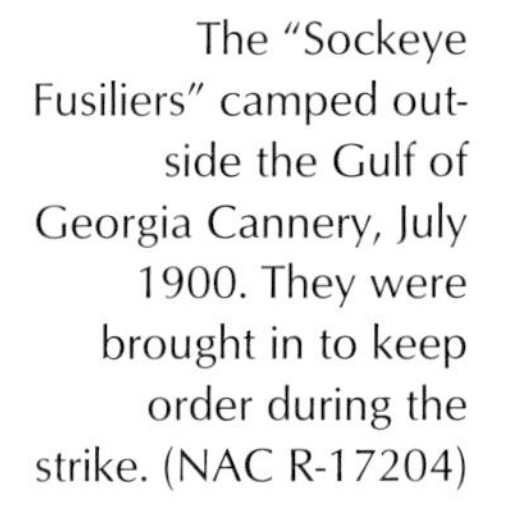
The "Sockeye Fusiliers" camped outside the Gulf of Georgia Cannery, July 1900. They were brought in to keep order during the strike. (NAC R-17204)

The Duke of Connaught's Own Rifles, known as the "Sockeye Fusiliers," 1900. (CVA Mil P39 N61)

dike, where Chief Constable Shea was halted by a sentry at Malcolm and Windsor's cannery. The constable indignantly protested that he was not to be interfered with by any common militiaman, and insisted that the dike being municipal property, it was his duty and privilege to keep it clear for the public. After a hot argument as to jurisdiction over the dike, the constable grabbed the bayonet of the soldier, who regained it by twisting—to the pain as well as the discomfiture of the policeman—and then gently prodded the officer with its point, the sergeant being at the same time called upon. He heard the two sides of the case and, of course, decided the soldier to be in the right, and that the military and specials were supreme upon the dike. (Vancouver *Province,* July 30, 1900)

The following season, 1901, the canners offered 12.5 cents per fish until the peak of the run had passed, then 10 cents per fish. The fishermen wanted fifteen cents all season. This was the year the province introduced a bill to take control of the fisheries, with a provision to create and auction off trap licences. Traps could supply fish at fifteen cents per piece with much less manpower, and fishermen suspected that if the canners could rely on trap-caught fish, fishing jobs would disappear. Several Native groups met at Chilliwack in early June and voted to refuse the canners' offer. The union took up the issue of naturalization fraud because it believed that large-scale Japanese immigration at the time was an attempt to build up a cheap, nonunion workforce. When the strike began, the canners put pressure on their Japanese labour contractors who already had received pre-season advances, and some Japanese crews were sent fishing. After two crews were badly beaten, armed company patrol boats

Gillnetters unloading their catch to a steam packer. (BCARS 52934)

accompanied the Japanese fishermen, followed by union boats with "U" emblazoned on their sails, which picketed the fishing grounds. Rumours that canners were arming the Japanese led to threats of arming the pickets. Six union men were arrested for intimidation and had their weapons confiscated. Violence increased. Fishermen were put ashore and their boats set adrift; nets were destroyed. More weaponry was packed aboard fishing vessels. Union leaders were jailed and eventually the union settled for the canners' terms. When the big salmon run hit, crews could not keep up with the vast quantities of fish, and hundreds of thousands of unprocessed salmon were dumped into the river.

The unions had little bargaining power in the "off" years, though some job actions were conducted. In anticipation of the 1913 season—another "big" year, in fact the biggest year in the fishery's history—there was another strike. By this time the Japanese held nearly half the fishing licences and the Steveston Japanese Association accepted the canners' price of twenty-five cents per fish. Soon after, the Industrial Workers of the World began a leaflet campaign threatening strikes if the price was dropped. The IWW reached out to fishermen and shoreworkers, men and women, and all nationalities, urging them to stand firm for the wages they wanted. When the big run appeared, the canners promptly dropped their price per fish to fifteen cents and the fishermen, led by the Japanese, refused to fish. When the canners tried to bring in trap fish, shoreworkers refused to process them. The women shoreworkers made demands for twenty-five cents per hour, and time and a half overtime. The strike fell apart as the fish continued up river. White fishermen in New Westminster kept fishing and the Japanese in Steveston eventually accepted twenty cents per fish, while Native and white fishermen in Steveston continued striking. Native women, however, returned to work in the canneries while Japanese women remained out. Again, the abundance of fish overwhelmed the cannery capacity.

A good catch of salmon. (BCP)

The price plummeted, surplus fish was given away, and thousands of fish were again dumped into the river.

The 1913 season was the last big year for a long time. Railway construction in the Fraser Canyon caused a blockage at Hell's Gate, preventing the salmon from reaching their spawning grounds; it took decades to restore the abundant runs.

Early attempts at unionization featured groups with narrow and exclusive aims, focussed on maintaining their special privileges rather than uniting all fishery workers. The first coast-wide fishermen's union was the Grand Lodge of BC Fishermen's Union formed in September 1900. It had locals from the Fraser in the south to Port Simpson in the north, including a local of Cowichan Indians, but it dissolved in 1903. Between the turn of the century and 1935, a number of labour organizations were formed, most along the lines of a craft union based on fishing gear types such as gillnet, seine or troll. More often than not, unions were further divided by race, resulting in Native, Japanese and white gillnetter organizations. The major organizations by the mid-1930s were the BC Fisherman's Protective Association (BCFPA), established in 1919; the Upper Fraser Fisherman's Protective Association (UFFPA), which merged with the BCFPA in 1937; the Amalgamated Association of Japanese Fishermen, established in 1926; the United Fishermen's Federal Union #44 (1932); the Native Brotherhood of British Columbia, formed in 1931; and the Pacific Coast Native Fishermen's Association, established in 1936, which joined the Native Brotherhood in 1943.

The BCFPA and the UFFPA members

Fish collectors with deckloads of salmon. (Williams Bros., CVA 586/5588)

Gillnetter unloading to a collector, 1940s. (CVA)

were mainly white gillnetters who formed their organizations largely to agitate against the Japanese. Membership was restricted to whites and Natives, and members were on the executive committee of the White Canada Association, the most dogged antagonist of Japanese Canadians. The policy of these fishermen's associations was a narrow business unionism characterized by opposition to Japanese and to other fishing gear type uses, especially seiners. The anti-Japanese sentiment came from economic competition—the Japanese fishermen were blamed for breaking the strikes of the early 1900s. The Japanese fishermen had demonstrated their ability to outfish both whites and Natives and between 1900 and 1919 held close to 50 percent of all licences on the coast.

After 1919 the federal government implemented a policy of reducing the number of Japanese fishing licences and of licensing by areas. The Amalgamated Association of Japanese Fishermen was founded in 1926 as a direct result of licence reductions to the Japanese, and had the avowed aim of combating discrimination against Japanese in the fishing industry.

The United Fishermen's Federal Union (UFFU) was founded in 1932 to represent fishermen with purse-seine gear, but it later extended its membership to all fishermen in any part of the industry. That same year it secured a charter from the Trades and Labour Congress as local #44, with official jurisdiction over all fishermen in British Columbia. The next year, the UFFU failed miserably in organizing a strike of halibut fishermen, undermining the seine fishermen's confidence in its stability and value. By the end of the fishing season, most of the seiners deserted the UFFU in favour of the newly established and militant Fishermen's and Cannery Workers Industrial Union.

The Native Brotherhood of British Columbia was established in 1931 with the aim of furthering the general welfare of Natives. From the 1930s to 1942 its power base was centred in the north coast and its leadership was drawn mostly from northern Natives, especially Tsimshians. Before the 1950s northern members of the Brotherhood were bitterly antagonistic towards Japanese fishermen who "invaded" their fishing areas. The Japanese were basically small boat gillnetters, as were the northern Natives, so they competed for the same fishing stocks. The antagonism had more than a racial basis; there had always been conflict between northern and southern fishermen of all races when either group moved into the other's fishing area. This protective attitude to a local fishery was not only a north–south phenomenon; it also existed between residents of the same area. Lower Fraser River gillnet fishermen, for instance, did not welcome upper river gillnetters on their fishing grounds at the river's mouth.

While northern Natives were, in general, anti-Japanese due to fear of economic competition, southern Natives were supportive of the Japanese gillnet fishermen but antagonistic toward Yugoslav seiners. Southern Natives were predominantly seiners and thus competed with the Yugoslav seine fishermen. This gear conflict was one reason the southern Natives didn't join the Native Brotherhood until 1942, forming instead their own organization, the Pacific Coast Native Fishermen's Association (PCNFA) in the fall of 1936 after a devastating strike at Rivers Inlet in July 1936. In the strike, southern Natives, mainly Kwakiutl from Alert Bay and Cape Mudge, supported the strike along with the Japanese fishermen, but members of the Native Brotherhood and white fishermen from the northern area ignored the pickets, breaking the strike.

The southern Native seiners' animosity to their Yugoslav competitors was not as aggressive as that of the northern Natives towards the Japanese. This was evident in the 1938 seine fishermen's strike that resulted in the first union agreement for seine fishermen. The PCNFA agreed to support the Salmon Purse Seiners Union (SPSU), the Yugoslav seine fleet, by not fishing. Their aggressive negotiations played a large part in the establishment of the SPSU, the driving force behind the present-day UFAWU. This alliance was due to a common desire to raise the price of seine fish, not a general agreement on each other's aims, and the next year the Native group attempted legal action to stop the SPSU's drive to sign up Native members.

In 1942 the PCNFA became a branch of the Native Brotherhood to allow the two Native fishing unions to work together on their common problems. The removal of the Japanese Canadians from the coast in 1942 eliminated the Native fishermen's principal competitors. At the Native Brotherhood Convention in 1942, it was agreed that people of Japanese descent should be sent to Japan after the war, a stand which appears to have come principally from the northern members of the Brotherhood. In 1943 the Brotherhood was officially recognized by the British Columbia Department of Labour as the bargaining agent for Indian fishermen, thus becoming a competitor to the fishermen's union. Between 1943 and the 1950s, however, its strength as a bargaining agent came from close co-operation with the fishermen's union that had jurisdiction over most fishermen and shoreworkers. In fact, many Native fishermen and shoreworkers, especially in the northern coastal area, voluntarily

Fish packer entering the Fraser River with one of the largest loads of salmon ever. (BCP)

Unloading herring seine at the Gulf of Georgia seine loft. (DS)

The Gulf of Georgia seine loft crew, c. 1949. Back row: Pearl Lumley, Olaf Anderson, Emily Guernsey, Pete Horsefield, George Gardiner, Johann Strand, Conrad Pearson. Fourth Row: Billy Bain, Helen Trotter, Kay Wilson, Sally Gabriel, Malcolm Campbell (behind), Ramona McDonald, Mary Larson, Jackie Therrien, Agnes Tellefeur. Third row: Alec Sutherland, Matt Stenfanich, Mrs Camerer (at right). Second row: Alec Main, Carl Ulvestad, Madelaine Therrien, Sonia Jovich, Rose Godmaire, Walter Lee, Zeena Kulba. Among those in the front row, according to the photograph notes, are Roger Goodall, Charlie Tubs, Dave Main, Gordon Johnston and Walter Rasch (GGCS PC 92-0729-179-P)

Dick Jack aboard the Canadian Fishing Company vessel *Kitlup* (CVA)

joined the UFAWU. The Brotherhood resisted being absorbed by the militant UFAWU where Natives would be a minority group. Traditionally, Native fishermen were militant; but by 1941 their most powerful representative, the Brotherhood, was becoming more conservative in its bargaining approach, while the UFAWU maintained its militant, direct action approach.

Communist organizers played key roles in the industry-wide unification of fishery workers, fishermen, tendermen and shoreworkers, beginning in the early 1930s. The Fishermen's and Cannery Workers Industrial Union (FCWIU), founded in 1934, was opposed to the other business, craft and racially based labour organizations. This was a militant "red" union, like its predecessor, the Fishermen's Industrial Union (FIU), founded in 1931. The FIU and the FCWIU were the first revolutionary unions to penetrate the North American fishing industry. Both were directly affiliated with the Communist Party through the Workers Unity League (WUL), established in conformity with the 1928 decision of the Sixth Congress of the Third International. In September 1931, a group of fishermen in Vancouver, aided by WUL organizers, formed the FIU of Canada. This new organization announced that:

> **it takes into the ranks all seine-boat fishermen, gillnetters, trollers, tender workers, cannery workers, saltery workers, fish reduction workers, and all workers employed in the fishing industry, regardless of sex or nationality. Our program is for organization of all workers employed in the fishing industry in Canadian waters into a militant industrial union affiliated to the Worker's Unity League and firmly linked with the Revolutional Trades Unions of the World.**

By 1933 the FIU had considerably increased its membership, but had failed to establish a unified front in the BC fishery. It was thus reorganized into the FCWIU whose aims were to support the strategy and tactics of revolutionary class struggle as outlined by the WUL: to organize all workers in the fishing industry into one Industrial Union and to rely entirely upon the militant activities of the organized fishermen and workers employed in the fishing industry. The FCWIU engineered a series of local "job action" strikes which raised fish prices substantially for both gillnetters and seiners. These disputes were bitterly fought and at the end of the season, opinion was divided as to whether fishermen had benefitted from them.

The United Fishermen and Allied Workers Union (UFAWU) was formed in 1945 by amalgamating the United Fishermen's Federal Union, representing BC fishermen, and the Fish, Cannery, and Reduction Plant and Allied Workers Union, representing BC shoreworkers. Uniting maritime workers and shoreworkers into a single industrial organization was a triumph of labour organizing; the fishery had seen unions since the 1890s, but all had failed in the quest for industrial unity before the formation of the UFAWU.

The work of the union movement made great changes in the lives of fishermen and cannery workers in Steveston. Working conditions, hours and wages were standardized and the Chinese contract system was abandoned. The union was instrumental in helping the returning Japanese settle back into the community. The camaraderie engendered by the countless meetings and work actions required to form the union left its mark on the social development of Steveston. The development of strong union leadership spilled over into other aspects of community life, creating institutions such as credit unions, and strong political influences. The union helped heal the wounds inflicted on Steveston during World War Two and laid the groundwork for a postwar community that could integrate all elements of its past to build its future.

Chapter Five:
Conclusion: The Phoenix

Downtown Steveston. (VPL 47)

Steveston: the name evokes a myriad of images—the proud silhouette of netlofts along the Fraser, the quiet bustle of Moncton Street, the colourful throngs of fishermen crowding the docks in summer months. Perhaps the word "tradition" best describes the spirit of Steveston, for the village has a uniqueness and heritage like no other community in the Lower Mainland. Steveston takes its history seriously. A cultural mosaic of Japanese, Chinese, Indian, and Anglo-Canadian has vivified its industry, business, and community for over one hundred years...fishing the waters of the Fraser, packing salmon in the canneries, harvesting the fertile farmlands, and developing a unique community solidarity which has kept it apart from the shifting suburbs of Richmond.

Steveston's physical appearance balances the small stores on Moncton Street and the neat bungalows and manicured Japanese gardens of the town with the haunting prospect of abandoned canneries less than a mile away. The waterfront is the old amongst the new...Now, as at the turn of the century, Steveston holds the distinction of being the largest commercial fishing harbour in Canada. This industry has both physically and spiritually defined the essential character of the town. (Victoria Kendall, *The Spirit of Steveston,* 1986)

Pioneers were not the only ones to shape Steveston's history. Natural disasters such as floods, storms and fires also have a place in the town's annals. Flooding was a yearly threat during the Fraser River's spring freshet. Garry Point Cannery, built in 1889, was abandoned in June 1893, as seventy feet of wharf and buildings dropped into the river. The most serious flooding occurred in 1894 when the water upriver in the Hell's Gate canyon rose ninety feet above its lowest known level. Downriver, at points jutting out along the shore of Lulu Island, canneries were swept away. By the time the freshet had lessened and the flood receded, five thousand cases of empty tins had disappeared and two-thirds of the cannery situated on Holly Point was a twisted wreck.

At 5 A.M. one June morning in 1894 I was a passenger on this boat and as we left the wharf at Steveston and made for the Westham Island shore we observed a long white line extending out to where the river flow entered the Gulf of Georgia, and, looking up river, stretching as far as the eye could see. On closer approach the white line proved to be a procession of bright, shiny, empty salmon tins, minus tops, following one behind another down river and bobbing up and down as the chop of the sea agitated them. Occasionally a can would tip over too far and would ship some water. This gradually lowered its level in the stream and subsequent additions to its liquid contents would result in its being completely filled, when it sank from sight. Other tins appeared to advance to close the gap and the line continued to hold its unity and close formation. (Henry Doyle)

Waves washing over the Steveston dyke and the Interurban track, 1951. (CRA 1977-1-24)

This flood was strong enough to change the course of the river, scouring a new channel.

Fires, too, had a major impact. Cannery buildings were volatile, especially when they were as closely packed together as they were along Steveston's waterfront. Fire insurance regulations required whitewashed cannery buildings and corrugated iron over the wooden shingle roofs as fire retardants, but when put to the test these measures did little to forestall destruction. Star Cannery burned in 1895.

When the nets were in the net loft and they had put the bluestone on them, and were not quite through at the cannery, a pile driver was driving piles, and a spark from the pile driver went into the loft and started a fire in the middle of the afternoon. This was about May of 1895...Right across from our house the men's house burned down, the fishermen's bunkhouse. The Japanese had a boat house at the end of 3rd Avenue. These Japanese poured water on their boathouse and kept it wet all the time while the wind was that way from the southeast and blew the flames against the boat house. Then the wind changed...They began to rebuild almost before the ashes were cold. We were wakened up at 4:00 in the morning by hammering putting shingles on the roof. (Ida Steeves)

In 1897, a petition was presented to the council signed by the property owners of Steveston representing more than half of the value of the real property within the said town-

Storm damage, 1951. (CRA 1977-1-25)

site, asking that a bylaw be passed to authorize the levying of a special rate on the property within the said townsite, to raise the sum of $1000 for the purpose of procuring means to extinguish fires; the said sum to be raised in one year. In response to this petition a bylaw was passed for the purpose asked for, under the name of the "Steveston Fire Department Bylaw." And in May of that year another bylaw was passed in this connection, setting forth the manner in which the committee of wardens should be appointed and the powers given them to carry out the provisions of the bylaw. This shows that the people of Steveston were conscious of the danger of fire to their town and were willing to pay for the protection they asked for. (Thomas Kidd)

Stevestonites were undoubtedly conscious of the great losses caused by fires in New Westminster and Vancouver, and of having millions of dollars of investment sitting in highly flammable canneries at their very doorstep. When major fires occurred, help was needed from neighbouring municipalities, and after the Steveston volunteer fire brigade was disbanded, fire-fighting equipment came from Vancouver to battle any blazes.

In October 1908, a large portion of the eastern part of the community, mainly Chinese homes, went up in flames with a property loss esti-

Firefighters and equipment outside the Walker Emporium during the 1918 fire. (CRA 1977-23-2)

The Steveston waterfront after the 1918 fire. Three canneries were destroyed in the fire. The Gulf of Georgia Cannery is in the background. (GGCS)

Downtown Steveston razed by fire, probably in 1918. (CRA)

mated at $35,000. In 1917 another fire destroyed seven buildings, again in the Chinatown area. A particularly calamitous fire on May 14, 1918, caused by an upset lantern in a Chinese bunkhouse at Star Cannery, destroyed Star, Steveston and Lighthouse Canneries and fifteen to twenty other buildings.

> The fire of 1918, that was the one where the fire engine from Vancouver broke down before it got here and when they did get here—some of the Japanese had stores along there in the part where I said Chinatown used to be, one of them packed his safe out in the middle of the road and in the thick smoke the fire engine hit it. That finished it. The only place left on that side of the street was the brick building. Where the post office used to be there was a big hotel, the London Hotel, with a terrace veranda out in front, high up. There was a Chinese bunkhouse in behind it, and apparently they were gambling or something, one of their card games, somebody upset the coal oil lamp and started the fire. It just happened that morning there was a very heavy west wind blowing, going about twenty-five, thirty mile an hour. It just took right through, nothing could stop it. They even dynamited several houses trying to stop it. Tried to save the canneries, but the sparks caught the riverside and we didn't have a fireboat. If we'd had a fireboat, we would have saved two canneries anyway. One of them had just got all the fish packed down from the north and it was full of canned salmon. That was one we were trying to save. At that, they salvaged quite a bit

A young Japanese woman with her child, walking through the debris of the 1918 fire. (CRA 1977-11-12)

The 1918 fire.
(VPL 9151)

because the floor burned and it dropped through—it was cooked anyway you see, so as long as the cans didn't break I guess they got a little salvage out of it. (Harold Steves, Sr.)

In 1924 fire claimed London's Landing, its warehouse and adjacent building on No. 2 Road. Finally, in 1926, Steveston lost Beaver Cannery, Winch Cannery, bunkhouses for fishermen and cannery workers, nets and boats in another fiery disaster. The volunteer fire brigades could not begin to control the blazes, and most of the property lost was not rebuilt.

In the end the fires, the poor salmon runs, and the technological innovations that reduced the canners' labour needs meant that the great canning crews were no longer required, and Steveston's wild summers now lived only as an exercise of memory. The fishing industry carried on, but a series of poor salmon runs meant only a few canneries survived. The hotels remained, but there was only one saloon, in the Sockeye Hotel. World War One took away many men. Richmond rebuilt its Town Hall, the symbol of the city centre, at Brighouse. The wild frontier excitement was over.

The boom days may have ended, but the community continued to grow at a steady pace after the war. Starting around 1907 the Japanese fishermen had begun bringing their wives to Steveston. Many new families were started, boosting the town's population. A new brickyard added more employment. In 1929 the CPR built a dock at the south

One of the many parades in Steveston.
(CRA)

end of 7th Avenue and started a ferry service from Steveston to Sidney on Vancouver Island. During the 1930s and 1940s there was no money to build new bridges, but the internal transportation network on Lulu Island improved. No. 9 Road and No. 19 Road were designated arterial highways and renamed Steveston Highway and Westminster Highway respectively.

The depression of the 1930s brought its share of deprivation, but living in a rich farming district, Stevestonites didn't suffer from lack of food. Many of the farms were reduced in size as farmers sold off some of their holdings. There were cutbacks in municipal services, but when tax bills loomed larger than the family bank account, municipal relief projects, such as dyke maintenance, helped out. Children got work picking strawberries and other produce, or were hired by a China contractor to watch machines or move cans. Most work was seasonal, either on the farms or in the canneries. Few new residents joined the community, although some migrants from the Prairies and eastern Canada worked at berry picking and other farm work. Holidays were usually spent at a cottage at Boundary Bay, Point Roberts or Birch Bay. Sports continued as a major leisure-time activity. During this time the *Richmond Review* started publication, providing Steveston residents with local news.

The Steveston Hotel, formerly the Sockeye Hotel, 1977. (DS)

Richmond maintained its own police force until 1942. It modernized its service with the purchase of a Moth aircraft, whose pilots surveyed flood damage and searched for missing boats, fugitives, rum runners, narcotics traffickers and smugglers. Eventually the Provincial Police took over responsibility for law enforcement.

Like so many others, Steveston farmers were caught by surprise when World War Two started, taking so much of the town's economic resources and labour. Shortly after Canada entered the war on September 10, 1939, Stevestonites started a civil defence fund; the local Japanese raised over $400 for this cause. By 1940 the town was involved in a full-scale war effort. Many of its young men enlisted, and citizens who were unable to go to the front joined the local militia or worked at the Boeing airplane or munitions plants in Richmond. Everyone helped to raise money, sell Victory Bonds, collect cans and magazines, and knit or sew for the cause. May Day celebrations were cancelled. There was a military presence in Steveston in the form of the 58th Battery with field guns on the Steves farm. The Pacific Coast Militia Rangers were on alert, and Stevestonites lived with air raid drills and blackouts. Food and fuel were rationed, the clocks were changed to "War Time" (Daylight Saving Time) to conserve energy, and there was no municipal spending on anything not directly related to the war effort. Leisure time was spent on civilian defence activities, Red Cross work and fundraising. Richmond police and fire departments assisted the military when needed, and the police used their Moth airplane to patrol the dykes and bridges. Steveston acquired a fire truck when the Steveston Air Raid Protection Unit (the first in Canada, ARP #1) decided to build one in case of an incendiary raid. They installed a two-ton Ford chassis in a truck body; the new creation had a pump and hose capable of pumping 150 gallons of water per minute and was the first mobile unit in Canada. The volunteer firefighters paid for the vehicle themselves, fearing they might have to move it elsewhere in Richmond if they accepted municipal money. Steveston merchants also built an ambulance.

Steveston was uniquely affected by the attack on Pearl Harbor in December 1941. Its Japanese citizens, the major part of its population, were suddenly under suspicion of being potential spies and saboteurs. The Richmond municipal council supported the federal government's decision in 1942 to uproot, dispossess and intern all West Coast residents of Japanese descent. In 1949, when the restrictions were finally lifted, some Japanese Canadians returned to the coast, but many eventually settled in other parts of Canada or went to Japan.

Mrs. Kumae Miyai, outside the Japanese housing at the Scottish Canadian Cannery, early 1940s. She and her husband and seven children were sent to a sugar beet farm during the internment. (NAC 92-0810-24)

World War Two revived the fishing industry. Canned salmon and herring were in demand to feed the troops. The few remaining canneries put out bigger packs and the industry grew again after many years of decline. Perhaps the first canner in Steveston was guided by foresight in naming his operation Phoenix.

After the war, as servicemen returned, found jobs and started families, Richmond developed into a suburb of Vancouver. It was close to the larger city and property was much cheaper. As the subdivisions went up, agricultural land disappeared and suburban amenities were put in place. Sewers and more roads were installed, and the Oak Street, Knight Street and Arthur Laing bridges and the Deas Island Tunnel made it easier to get into and out of Richmond by car. By 1957 the interurban was no longer needed to transport Stevestonites into Vancouver. The transportation links spurred internal growth, and there was a great surge in the construction of houses, light industry, warehouses, malls, shops, theatres, parks and recreational facilities such as the Minoru Pavilion, Ice Arena, Aquatic Centre, Arts Centre and Public Library. More churches, a hospital and other medical/dental and social services were needed and built. The development of the airport on Sea Island created many jobs for Richmond residents. Numerous community groups and service organizations developed.

Amidst all this growth, Steveston residents were determined to preserve their unique community. To this end they founded the Steveston Community Society in 1944. The Society has been involved in civic affairs and a leader in area park development and family-oriented recreation ever since. One of its first acts was to campaign for a junior playground on Broadway, since the Moncton Street Park was dominated by older children. They bought the land (earlier confiscated from Japanese residents) from the Custodian of Enemy Property, and celebrated the park's establishment with a dance at the Orange Hall on Steveston Highway. The sense of accomplishment led to new plans and a project was soon underway to landscape and equip Steveston Park on Moncton. As it was still wartime,

Last run of the Interurban (the tram from Vancouver to Steveston) in 1958. (CRA 1978-12-2)

Sophie Kuchma, the 1946 Salmon Queen, and her attendants. (CRA 1985-189-2)

the municipality was unwilling to contribute funds, and the society's membership fee of $1 was woefully inadequate, even after women were allowed to join. In the end members decided to do the work themselves, grading, seeding and pulling up blackberry bushes in work parties. Donations came in for playground equipment, and the society's vision grew more ambitious to include wading pools and benches. They planned a Sports Day on July 1, 1944 as a fundraiser, an event that turned into the annual Salmon Queen Carnival. The event was immensely successful, both in financial and morale-building terms, in part because Richmond residents missed the Dominion Day activities in Vancouver they had been forced to give up due to gasoline rationing. The society grew and involved itself in advertising, finance and civic works, and held another successful sports day in 1945. Plans for the 1946 event included hosting a Salmon Queen Contest. The first year's winner was the Princess (from a field of six, including the Canneries Princess, the Merchants Princess, the Brighouse Princess, the Bridgeport Princess, the Dairy Princess, and the East Richmond Princess) who sold the most tickets to the pageant. The contest generated enormous publicity and community excitement. July 1, 1946, was the first national holiday after the end of the war, and Vancouver put on a celebration designed to release a lot of pent-up festive spirit. The Society's Salmon Queen float in the

Mrs. Asae Murakami (fourth from left) with five children, near the strawberry fields of Kame Murakami, 1927. (BSHS)

Vancouver parade invited everyone back to Steveston for the Salmon Queen Carnival and many took up the offer. With more than two thousand in attendance, the event raised $3,500. The society also sponsored a Halloween bonfire, fireworks and costume parade for the community's children. In 1972 a Santa Claus Parade was added. The Salmon Queen festival continued to grow, adding a King of the Fraser competition, something of a fishermen's Olympics. The society went on to expand the original Steveston Park site to twenty-three acres and built the Steveston Community Centre, the Martial Arts Centre, swimming pools, sports facilities and more playgrounds and parkland.

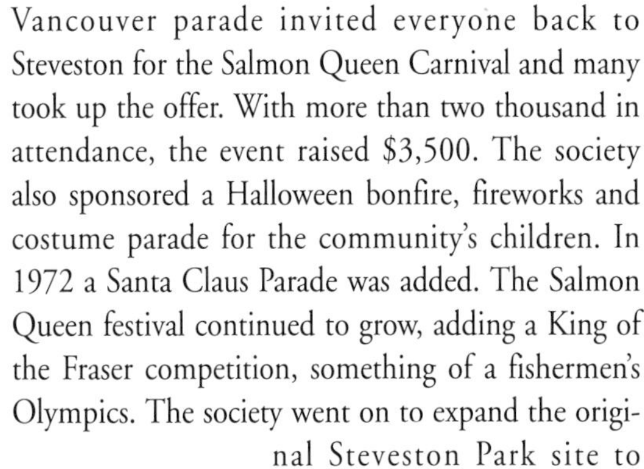

Peter Dignard (at centre) with Phyllis Johnson and Marg McDonald, outside the Moncton Coffee Shop. (GGCS)

The Steveston Historical Society has been active in preserving Steveston's historical character. The Steveston Museum, housed in the old Northern Bank building on Moncton Street, and the Britannia Heritage Shipyard, on the dyke between Trites Road and Railway Avenue, draw increasing numbers of tourists. The community hopes to use its heritage to boost the local economy. This trend continues with the development of a National Historic Site dedicated to the fishing industry at the old Gulf of Georgia cannery buildings at the foot of 4th Avenue.

The main building of the original complex was built in 1894 by the Gulf of Georgia Canning Company. Incorporating the latest technology available, it was one of the largest plants along the lower Fraser waterfront and was singled out by the Victoria *Colonist* as "the monster cannery at Steveston." Malcolm and Windsor bought the building in 1895 and added a second canning line, making this complex the second largest in BC. By 1902, when BC Packers Ltd. bought twenty-nine of the forty-nine plants along the Fraser River, the Gulf of Georgia plant was the largest in the province. BC Packers' consolidation led to fewer but larger plants; although the Gulf of Georgia expanded to three cannery lines in 1905, it was now the second-largest plant in BC.

The Canadian Fishing Company bought the site in 1926, taking over the cannery, net loft and fish-buying functions of the plant. The Depression brought a halt to cannery operations in 1930, although fish buying and net construction and repair continued. Only once after 1930, in 1946, did the Gulf of Georgia plant process salmon again. In 1940, however, wartime demand led to a reactivation of the plant for canning herring. At the same time, a herring reduction plant was set up to process the fish into oil and meal. Herring canning went on until 1947; the herring reduction process continued until 1979.

In 1977 the Small Craft Harbours Branch of the Canadian Department of Fisheries and Oceans acquired the Gulf of Georgia complex as part of its harbour redevelopment program at Steveston. It then leased the site back to the Canadian Fishing Company, primarily for use as a net loft where nets are constructed and repaired.

The Historic Sites and Monuments Board of Canada recommended the commemoration of Canada's West Coast fishing industry at Steveston in 1976. The Gulf of Georgia plant was suggested for preservation and interpretation because of its condition, accessibility and close association with the salmon and herring fisheries. Ownership was transferred to Environment Canada and its agency, Parks Canada, in 1984. The site is still in use as a net loft, remaining a vital part of the Steveston waterfront. A portion of the site was opened by the Parks Service in June 1994, as a national historic park.

Steveston uses history to nurture its community. It maintains its uniqueness in a landscape of suburbs by honouring the farming, fishing and multicultural heritage that gave it birth. Salmonopolis should endure well into the next century and stand as an example of a community that embraces the cultures and contributions of all its residents.

ENDNOTES

INTRODUCTION

The material in this chapter is from several sources, including Margaret Ormsby's *British Columbia: A History*, Leslie Ross's *Richmond: Child of the Fraser*, Thomas Kidd's *History of Richmond Municipality*, Daphne Marlatt's *Steveston Recollected*, and various interviews, tapes and summaries held by the Richmond City Archives.

CHAPTER 1

The information presented here is largely from the reminiscences of Ida Steeves, one of Manoah Steves's daughters. These records are in the Richmond City Archives. Thomas Kidd's history is also used, as is Leslie Ross's book. Records of the West Lulu Island Dyking Commission and many of the interviews in the Richmond City Archives are also incorporated, as are articles from the *Vancouver News* and the *Richmond Review*.

CHAPTER 2

This chapter makes extensive use of the memories of Ida Steeves and Thomas Kidd, as well as information in *Above the Sandheads*, the memories of T. Ellis Ladner as recorded by his daughter, Edna Ladner. The oral histories recorded in *Steveston Recollected* and the taped interviews and police reports in the Richmond City Archives provided much information. Articles from various Vancouver and Victoria newspapers, and letters from missionaries printed in *Work for the Far West*, *Missionary Bulletin* and *Missionary Outlook* yielded much detail of the life of the times. The letterbooks of David Lew, a Chinese contractor whose papers are held at the University of British Columbia Library, Special Collections, were also useful. David Jelliffe correlated the interviews of over one hundred Richmond pioneers to provide an account of Steveston life at the turn of the century as part of a Richmond Nature Park project.

CHAPTER 3

This chapter is comprised of information gleaned from W.A. Clemens and G.V. Wilby's *Fishes of the Pacific Coast of Canada*, Duncan Stacey's *Sockeye and Tinplate*, and Ladner's *Above the Sandheads*. Numerous interviews in the Richmond City Archives deal with this subject and the interviews in the Britannia Heritage Shipyard Oral History Project, *Steamboxes, Boardwalks, Belts and Ways*, were full of material. Again the *Missionary Outlook*, *Methodist Recorder* and *Missionary Bulletin* provided insights. The papers of Edwin DeBeck and Alfred Carmichael at the British Columbia Archives and Records Service are very descriptive. A great deal of material was also found in the Canadian royal commissions on the fishery and on Chinese immigration. The reports of the US fishery commissioner also contributed numerous facts. Various publications such as *The Canadian Fisherman*, the *Vancouver Daily World* and the *British Columbia Magazine* were also useful.

CHAPTER 4

This chapter relies heavily on *Sockeye and Tinplate* and on the publications Duncan Stacey wrote for Parks Canada's Gulf of Georgia Cannery project. It also makes extensive use of reports of fishery commissioners in BC and the US, sessional papers, and royal commissions. The works of Edna Ladner, Geoff Meggs (*Salmon: The Decline of the British Columbia Fishery*) and Ken Adachi (*The Enemy That Never Was*) were also used. The reminiscences found in *Steamboxes, Boardwalks, Belts and Ways* and in *Steveston Recollected* as well as interviews in the Richmond City Archives and the DeBeck papers, filled out the picture. Newspapers and magazines such as the Vancouver *Province*, *The Canadian Fisherman*, and *Missionary Outlook* also had useful information.

CHAPTER 5

This chapter draws on material in Leslie Ross's work, in Victoria Kendall's *The Spirit of Steveston*, and the Henry Doyle papers at the University of British Columbia Library, Special Collections. Ida Steeves's and Thomas Kidd's memoirs were also used.

BIBLIOGRAPHY

ARTICLES

Anderson, Alexander C. "The Dominion of the West." Government Prize Essay, 1872. Victoria BC: printed by Richard Wolfenden, Government Printer, 1872.

Cowan, John B. "The Queerest Town in Canada," in *Busy Man's Magazine* (March 1910).

Gladstone, Percy and Stuart Jamieson. "Unionism in the Fishing Industry of British Columbia," in *Canadian Journal of Economics and Political Science*, Vol. 16 (February 1950), pp. 1–11.

Gowen, Herbert. "Salmon Fishing and Canning on the Fraser," in *The Canadian Magazine* (1893), pp. 159–65.

Greenwood, W. "The Salmon Fishermen," in *Canadian Fisherman* (July 1917), pp. 288–94.

Hume, R.D. "The First Salmon Cannery," in *Pacific Fisherman* (January 1904), pp. 19–21.

"Our Salmon and Salmon Canneries," in *The Resources of British Columbia, Vol. I* (December 1883).

Ralston, H. Keith. "Patterns of Trade and Investment on the Pacific Coast, 1867–1892: The Case of the British Columbia Salmon Canning Industry," in *BC Studies*, No. 1 (winter 1968–69), pp. 37–45.

Reid, David. "Company Mergers in the Fraser River Salmon Canning Industry, 1885–1902," in *Canadian Historical Review*, Vol. 56 (1975).

Weston, Garnett. "Steveston-by-the-Fraser," in *British Columbia Magazine*, Vol. 7, No. 8 (August 1911).

BOOKS

Adachi, Ken. *The Enemy That Never Was: A History of the Japanese.* Toronto: McClelland and Stewart, 1976.

Carrothers, W. *The British Columbia Fisheries.* Toronto: University of Toronto Press, 1941.

Chapelle, Howard. *American Small Sailing Craft, Their Design, Development and Construction.* New York: W.W. Norton, 1951.

Clemens, W.A. and G.V. Wilby. *Fishes of the Pacific Coast of Canada.* Ottawa: Fisheries Research Board of Canada, 1961.

Cobb, John. *The Canning of Fishery Products.* Seattle: Miller Freeman, 1919.

_________. *Pacific Salmon Fisheries, 4th ed.* Washington: Government Printing Office, 1930.

Collins, J.W. "The Fishing Vessels and Boats of the Pacific Coast." *Bulletin of the United States Fish Commission*, Vol. X (1890). Washington: Government Printing Office, 1892.

Goode, George Brown. "The Salmon Fishing and Canning Interests of the Pacific Coast," in *The Fisheries and Fishery Industry of the United States, Vol. I, Sec. 5.* Washington: Government Printing Office, 1887.
Ham, Leonard C. *A Prehistoric Heritage Resource Overview of Richmond, B.C.* Richmond BC: Richmond Museum, Department of Leisure Services, Corporation of the Township of Richmond, 1986.
Henderson's British Columbia Directory and Street Index [name varies]. Victoria and Vancouver: Henderson Publishing Company, 1889–1910.
Kendall, Victoria. *The Spirit of Steveston: A History of the Steveston Community Society.* Steveston Community Society, 1986.
Kidd, Thomas. *History of Richmond Municipality,* Wrigley Printing Company Ltd., 1927. Reprinted by Richmond Printers Ltd., 1973.
Kipling, Rudyard. *From Sea to Sea: Letters of Travel, 2 vols.* New York: Doubleday and McClure, 1899.
Knight, Patricia. *Centennial History of St. Anne's Anglican Church—Steveston,* 1891–1991. St. Annes'–Steveston Trustees, 1991.
Knight, Rolf. *Indians at Work.* Vancouver: New Star Books, 1978.
Ladner, T. Ellis. *Above the Sandheads: A Vivid Account of Life on the Delta of the Fraser River.* Edna G. Ladner, 1979. Printed by D.W. Friesen and Sons, Cloverdale BC.
Lyons, Cicely. *Salmon: Our Heritage.* Vancouver: British Columbia Packers Ltd., 1969.
McEvoy, Bernard. *From the Great Lakes to the Wide West.* Toronto: William Briggs, 1902.
Marlatt, Daphne. *Steveston Recollected.* Victoria BC: Provincial Archives of British Columbia, Aural History, 1975.
Meggs, Geoff. *Salmon: The Decline of the British Columbia Fishery.* Vancouver: Douglas & McIntyre, 1991.
Meggs, Geoff and Duncan Stacey. *Cork Lines and Canning Lines.* Vancouver: Douglas & McIntyre, 1992.
Ormsby, Margaret A. *British Columbia: A History.* Toronto: Macmillan, 1971.
Rathbun, Richard. "A Review of the Fisheries in the Contiguous Waters of the State of Washington and British Columbia, 1899," in Commission of Fish and Fisheries Commissioners Report. Washington: Government Printing Office, 1899.
Ross, Leslie. *Richmond: Child of the Fraser.* Richmond BC: Richmond '79 Centennial Society, 1979.
Rounsefull, George and George Keleg. *The Salmon Fisheries of Swiftsure Bank, Puget Sound and the Fraser River.* Washington: Government Printing Office, 1938.
Stacey, Duncan. *Sockeye and Tinplate: Technological Change in the Fraser River Salmon Canning Industry, 1871–1912.* Heritage Record No. 15. Victoria BC: British Columbia Provincial Museum, 1982.
Steamboxes, Boardwalks, Belts and Ways: Stories from Britannia, Britannia Heritage Shipyard Oral History Project, Vol. I, 1992.

FEDERAL AND PROVINCIAL GOVERNMENT RECORDS

British Columbia. *British Columbia, Its Present Resources and Future Possibilities.* Victoria BC: The Colonist Printing and Publishing Company, 1893.
_________. *Sessional Papers.* Victoria BC: 1870–1914.
Canada. British Columbia Fishery Commission, "Report and Minutes of Evidence, 1892," *Sessional Papers,* 1893, no. 10c.
_________. Department of Marine and Fisheries [departmental name varies]. "Annual Reports," 1872–1914. *Sessional Papers,* 1873–1915.
_________. Department of Labour. *Labour Gazette,* 1900–1913.
_________. Dominion Fisheries Commission for British Columbia, 1905–1907. "Report and Recommendations." Ottawa: Government Printing Bureau, 1908.
_________. "Report of Special Fishery Commission, 1917." Ottawa: King's Printer, 1918.
_________. Royal Commission Appointed to Investigate Alleged Chinese Fraud and Opium Smuggling on the Pacific Coast, *Report,* 1910–1911.
_________. Royal Commission on Chinese and Japanese Immigration. "Report," 1902. *Sessional Papers,* 1902, no. 54.
_________. Royal Commission on Chinese Immigration, "Report and Evidence," *Sessional Papers,* 1885.
Reid, David. *The Development of the Fraser River Salmon Canning Industry, 1885–1913.* Department of the Environment, July 1973.

RICHMOND CITY ARCHIVES

Biography Files, Ida Steeves, as recorded and compiled by Great Cheverton and Kathy Steves.
Interview with Jack Downing, in *Draft for the Marine Development of Richmond; Based on Reminiscences of Present or One Time Richmond Residents, Tape Recorded, Transcribed, and Presented Here in the Form of a Personalized History.* Manuscript.
Inventory West Lulu Island Dyking Commission, (MA 16), November 1988, doc. 1626c.
Police Reports.
Reference Files, Richmond '79 Centennial Society, Historical Sub-Committee.
- MS 2, "Richmond's First Fishermen," by John Belshaw.
- MS 5, "Boozing It Up in the Good Old Days," by Ruth Leaming.
- MS 7, "The Rise and Fall of a Boomtown, Steveston, B.C.," by John Belshaw.
- MS 13, "Steveston–Gambler's Paradise," by Ruth Leaming.
- MS 21, "The Hong Wo Store–Richmond's First Supermarket," by John Belshaw.
- MS 30, "Potato Problems of the 1930's," by Barbara Hynek.
- MS 31, "A Day at the Races," by John Belshaw.
- MS 32, "Murder by Brush Hook," by Ruth Leaming.
- MS 36, "May Day in Richmond," by Barbara Hynek.
- MS 43, "The Fishing Strike of 1900," by Barbara Hynek.
- MS 50, "Early Agricultural Fairs," by Ruth Leaming.
- MS 51, "ByGone By-Laws," by Barbara Hynek.

Summaries of Interviews: Anderson, Anonymous fisherman, Atchison, Baldwin, Blair, Bothwell, Bussanich, Buchanan, Calvert, Capadouca, Cheverton, Christianson, Clark, Cotton, Deagle, Dick, Dobrilla, Dumont, Easthope, Elston, Flury, Frank, Fraser, Gillespie, Gilmore, Hall, Hayashi, Herrling, Howard, Hunter, Johnson, Kennedy, Kojima, Marshall, Matsuo, Matsuzaki, May, Mizuguchi, Montgomery, Moore, Moreside, Mort, Murakami, McClelland, MacKay, MacKenzie, Nelson, Okano, Paiger, Parkin, Phillips, Point, Ransford, Sakamoto, Shaw, Skinner, Smith, Sparrow, Steves, Thompson, York.

RICHMOND NATURE PARK

Richmond Nature Park, 11851 Westminster Highway, Richmond BC, Pioneer Historical Publications

- Publication No. 2, *Steveston, Boom Days–Fire,* by Dellis Cleland.
- Publication No. 4, *Childhood in the Old Days of Richmond,* by David Jelliffe.
- Publication No. 7, *Fishing the Fraser: Early Days.*

NEWSPAPERS AND PERIODICALS

British Columbia
British Columbian (New Westminster)
The Canadian Fisherman (Montreal), 1914–1930
Colonist (Victoria)
Columbian (New Westminster)
The Fisherman (Vancouver)
Mainland Guardian
Methodist Recorder
Missionary Bulletin
Missionary Outlook
Pacific Fisherman (Seattle)
Province (Vancouver)
Richmond Review
Vancouver Daily World
Vancouver News
West Coast Fisheries (San Pedro)
Western Fisheries (Vancouver)
Work for the Far West: A Quarterly Magazine for the Diocese of New Westminster

THESES, DISSERTATIONS, MANUSCRIPTS AND OTHER PAPERS

Anglo British Columbia Packing Company. Records and Papers, University of British Columbia Library, Special Collections.

Appleyard, Rev. *Missionary Work at Port Essington.* Manuscript, British Columbia Archives and Records Service.

Bell-Irving, Henry. Diaries, Vancouver City Archives.

"British Columbia Salmon Canneries." Typescript compiled by H. Keith Ralston, 1965. British Columbia Archives and Records Service.

Carmichael, Alfred. *Account of a Season's Work at a Salmon Cannery; Windsor Cannery, Aberdeen, Skeena.* Manuscript, British Columbia Archives and Records Service.

Carmichael, Alfred. *The Emigrant Boy.* Manuscript, British Columbia Archives and Records Service.

"Correspondence from Canoe Pass Cannery, 1892–1896." Manuscript, University of British Columbia Library, Special Collections.

DeBeck, Edwin Keary. Papers, British Columbia Archives and Records Service, add. mss 346.

Deutsch, J. et al. "The Fishing Industry of British Columbia," in *Economics of Primary Production in British Columbia, Vol. III, 1959.* University of British Columbia Library, Special Collections.

Doyle, Henry. "Rise and Decline of the Pacific Salmon Fisheries," 2 vols. Manuscript, University of British Columbia.

_________. Papers, University of British Columbia Library, Special Collections.

"Fish Cleaning Machines; References for Smith Investigation." Archives manuscript, Shiels Papers, Western Washington State College, Geography Department.

Fort Langley Journal. International Pacific Salmon Fisheries Commission.

Gladstone, Percy. "Industrial Disputes in the Commercial Fisheries of British Columbia." MA thesis, University of British Columbia, 1959.

J. H. Todd Company. Records and papers, University of British Columbia Library, Special Collections.

Lawrence, Joseph. "An Historical Account of the Early Salmon Canning Industry in British Columbia, 1870–1900." Graduating essay, University of British Columbia, 1951.

Lew, David C. *Letterbook, June 1907–September 1910.* Manuscript, University of British Columbia Library, Special Collections.

Officer, Ernest. "British Columbia's Salmon Industry." Graduating essay, University of British Columbia, 1955.

Ralston, Keith. "The 1900 Strike of Fraser River Sockeye Salmon Fishermen." MA thesis, University of British Columbia, 1965.

Simpson, McTavish. "Fifty Years Ago in the Canning Industry." Essay, April 17, 1941, British Columbia Archives and Records Service.

Somers, Diane. "W.H. Steves and Steveston: A Case Study in Pioneer Capitalism," for Victoria City Archives.

Strong, Gordon. "The Salmon Canning Industry in British Columbia." Graduating essay, University of British Columbia, 1934.

INDEX